U0155942

[美国]劳伦斯·普林西比 著　张卜天 译

科学革命

牛津通识读本·

The Scientific Revolution

A Very Short Introduction

译林出版社

图书在版编目（CIP）数据

科学革命／（美）普林西比（Principe, L.）著，张卜天译．—南京：译林出版社，2013.12（2024.10重印）
（牛津通识读本）
书名原文：The Scientific Revolution: A Very Short Introduction
ISBN 978-7-5447-4518-5

Ⅰ.①科… Ⅱ.①普… ②张… Ⅲ.①自然科学史－世界 Ⅳ.①N091

中国版本图书馆 CIP 数据核字（2013）第 242898 号

著作权合同登记号　图字：10-2023-121 号

科学革命［美国］劳伦斯·普林西比／著　张卜天／译

责任编辑　何本国
责任印制　董　虎

原文出版　Oxford University Press, 2011
出版发行　译林出版社
地　　址　南京市湖南路 1 号 A 楼
邮　　箱　yilin@yilin.com
网　　址　www.yilin.com
市场热线　025-86633278
排　　版　南京展望文化发展有限公司
印　　刷　江苏凤凰通达印刷有限公司
开　　本　890 毫米 × 1260 毫米　1/32
印　　张　9.125
插　　页　4
版　　次　2013 年 12 月第 1 版
印　　次　2024 年 10 月第 12 次印刷
书　　号　ISBN 978-7-5447-4518-5
定　　价　39.00 元

序言

吴国盛

自柯瓦雷创造"科学革命"这个概念以概括16、17世纪欧洲思想所发生的激进变革以来，它成了科学史家们最热切关注的中心话题。围绕这个话题，70年来产生了多种不同的编史纲领、数以百计的专门著作和几代优秀的科学史家。事实上，正是这个主题培育并造就了科学史学科的独特范式。我们或许可以说，时至今日，一个不深入了解"科学革命"的科学史家，不是一个合格的科学史家；一个科学史的入门初学者，最好是先读关于"科学革命"的著作。

然而，"科学革命"主题已经严重发散，正像本书作者在引言中所说："如果问10位科学史家科学革命的实质、时间段和影响是什么，你可能会得到15种回答。"甚至，像夏平这样的科学史家根本否认存在什么"科学革命"。在这种众说纷纭、莫衷一是的情况下，对专业科学史家来说，应该去比较各家的观点和论据，去研究关于科学革命的种种编史学；而对初学者来说，则需要一本权威性的综合著作，总结各方观点的合理之处，讲述一个主题鲜明的历史故事。所幸的是，本书就是这样的著作。

作者劳伦斯·普林西比（Lawrence Principe）是约翰斯·霍普金斯大学科学技术史系和化学系双聘讲席教授，1983年毕业于特拉华大学，获化学和文科双学士学位，1988年获印第安纳大学有机化学博士学位，同年进入约翰斯·霍普金斯大学化学系任教，1996年获约翰斯·霍普金斯大学科学史博士学位。他的主要研究方向是近代早期化学史，特别是炼金术与化学的关系史。普林西比不仅是一位优秀的职业科学史家，而且还是一位杰出的教师，很擅长用简明通俗的语言条理分明地讲述历史。他曾获得卡内基基金会、邓普顿基金会以及约翰斯·霍普金斯大学多次颁发的教学大奖。美国教学公司（The Teaching Company）从上世纪90年代开始组织全美名师录制“名课”（The Great Courses），其中的“从古代到1700年的科学史”（History of Science: Antiquity to 1700）即由普林西比讲授。

按照天界、地界、生命界、人工界的顺序，本书既讨论了天文学—力学—物理学这个科学革命叙事的传统线索，也讨论了占星术—炼金术—赫尔墨斯主义等化学论的叙事线索，还把解剖学—医学—植物和动物博物学也纳入科学革命的范畴中。这种种线索的并存并没有损害全书鲜明的主题。作者把科学革命时期（大约从1500—1700年）通过错综复杂的连续渐变造成的最大断裂总结为，一个处处关联的、充满意义的、隐含神圣设计和无声隐喻的世界被彻底瓦解，具有宽广视野和多面经验的自然哲学家，被专业化、分科化的技术科学家所取代。革命之前的“关联宇宙论”一直或多或少地存在于革命的全过程中，它可以帮助我们理解为什么开普勒、玻意耳、牛顿等人始终坚持认识自然就是认识上帝，自然哲学是神学的分支，宗教才是研究自然的原动力；为什么牛顿会热衷于研究炼金术和圣经年代学，并且相信摩西等古代哲人早就知道万有引力定律；为什么第谷在天堡里同时进行天文观测和炼金术（他称之为“地界天文学”）的工作。甚至机械自然观的出现，也应该从“关联宇宙论”与人类日益增加的技术能力的共同背景下加以理解。

以寥寥数万字的篇幅，把近几十年发掘出来的如此之多的线索组织起来，并且提炼出鲜明的科学革命形象，这是本书的最大长处和特色。我相信，多年来深受实证主义及辉格式科学史影响的中国读者，读读本书，会有眼前一亮的感觉。

目录

致谢 I

引言 I

1 新世界和旧世界 1

2 关联的世界 16

3 月上世界 32

4 月下世界 57

5 小宇宙和生命世界 81

6 科学世界的建立 99

尾声 117

索引 120

英文原文 133

致谢

感谢一些朋友和同事阅读了本书的全部或部分内容并且提出了批评，还有人与我讨论了如何才能把科学革命压缩成如此简洁的形式，特别是帕特里克·J.博纳、H.弗洛里斯·科恩、K.D.孔茨、玛格丽特·J.奥斯勒、吉安娜·波马塔、玛丽亚·波图翁多、迈克尔·尚克和詹姆斯·弗尔克尔。还要感谢为本书提供图片的人，他们是：化学遗产基金会的詹姆斯·弗尔克尔；约翰斯·霍普金斯大学谢里登图书馆特藏部的厄尔·黑文斯及其同事们；康奈尔大学克罗赫图书馆珍本手稿特藏部的大卫·W.科森及其同事们。

尤其怀念与我的同事和朋友玛吉·奥斯勒[①]的多次交谈。我们边喝单一麦芽苏格兰威士忌，边讨论如何编写近代早期科学史。她的过早离世使世界变得更加贫乏和无趣。谨以此书纪念她。

① 即玛格丽特·J.奥斯勒。——译注

引言

1664年底，天空中出现了一颗明亮的彗星。西班牙人最先注意到了它，但接下来几周，这颗彗星变得越来越大、越来越亮，全欧洲都把目光投向了这一天象奇观。在意大利、法国、德国、英格兰、荷兰等地，甚至是欧洲在美洲和亚洲新近占领的殖民地和偏远地区，观测者们都在追踪和记录这颗彗星的运动和变化。一些人做了认真测量，争论着彗星的大小和距离以及在天空中的轨道是直还是曲。一些人用肉眼观察它，另一些人则用刚刚问世60年左右的望远镜之类的仪器进行观测。一些人试图预言它对地球、天气、空气质量、人的健康、人类事务和国家命运的影响。一些人视之为检验新天文学思想的良机，另一些人则视之为神的预兆（不论好坏）。印刷的小册子层出不穷，新的自然现象类期刊杂志刊登了论文和争论，人们在宫廷和学院、咖啡馆和小酒馆讨论它，相距遥远的观察者频繁通信，交换着丰富的思想和数据，编织出超越政治和信仰的交流网络。全欧洲都在注视着这一自然奇观，力图理解它并从中受益。

1664至1665年的这颗彗星仅仅是一个例子，表明17世纪的欧洲人不仅交流密切，而且密切关注他们周围的自然界并与之互动。透过不断改进的望远镜，他们看到了广袤的新世界——意想不到的木星卫星、土星光环和无数新的恒星。透过同样新近发明的显微镜，他们看到了蜜蜂螫针的精细结构、放大到狗的尺寸的跳蚤，发现醋、血液、水和精液中居然还存在着一群从未想到的“微动物”。利用解剖刀，他们揭示出植物、动物和人的内部运作方式；借助火，他们把自然物分解成化学组分，将已知物质结合成新物质；依靠船舶，他们驶向新的陆地，带回关于新的植物、动物、矿物和民族的新奇样本和报告。他们设计出新体系来解释和组织世界，复兴古代体系，就彼此的优势展开无休止的争论。他们寻找隐藏在世界背后的原因、意义和寓意，追溯上帝的创造与维持之手的踪迹，试图借助新技术和隐秘的古代知识来控制、改进和开发他们所遭遇的世界。

科学革命——大约从1500年到1700年——是科学史上讨论最多的、最重要的时期。如果问10位科学史家科学革命的实质、时间段和影响是什么，你可能会得到15种回答。一些人把科学革命看成与中世纪世界的截然断裂，正是在科学革命时期，我们所有人（至少是欧洲人）变成了“现代的”。在这种观点看来，16、17世纪的确是革命性的。另一些人则试图把科学革命变成一个无效的事件，仅仅将其视为回顾历史时所产生的一种幻觉。不过，如今更多谨慎的学者认识到，虽然中世纪与科学革命之间存在着许多重要的连续性，但这并不能否认16、17世纪以令人震惊的重要方式利用和改造了中世纪的遗产。事实上，“科学革命”（现在更多被称为“近代早期”）**兼具**连续与变化的特征。这一时期就自然界发问的人明显增多，他们设计了新的途径对这些问题给出了大量新的回答。本书描述了近代早期思想家对周围世界的设想、研究、发现以及这一切对他们的意义，讨论了他们如何为近代科学的知识和方法奠定了基础，如何努力解决至今仍然困扰我们的问题，如何精心打造了充满美和希望的丰富世界，这样的世界我们常常忘记如何去观察。

第一章

新世界和旧世界

近代早期的成就奠基于中世纪建立的思想和制度。近代早期的人试图回答的许多问题都是在中世纪提出的，而用于回答它们的许多方法也源自中世纪的研究者。然而，近代早期的学者却乐于诋毁中世纪，宣称自己的工作是全新的，尽管他们保留和依赖的东西至少与抛弃或因时修改的东西同样多。从中世纪到近代早期的特定变化并非在整个欧洲同时发生，不论这些变化是思想的、技术的、社会的还是政治的。相比于英格兰这样的欧洲边缘地区，医学、工程、文学、艺术、经济和民事等明显的“现代”产物早已在意大利完全确立。同样，不同学科在不同时间出现了不同速度的发展。大约1500年到1700年这一时期——不论如何称呼它——是一幅观念和潮流的织锦，一个充斥着相互竞争的体系和概念的喧闹市场，一个涵盖了一切思想实践领域的忙碌的实验室。这一时期不断出现的文本表明了其作者对自己所处时代的异常兴奋。单凭一个标签、一本书、一位学者和一代人不可能理解它的全貌。为了理解这一时期及其重要性，我们需要近距离考察当时究竟发生了什么以及原因如何。

理解科学革命首先要知道它在中世纪和文艺复兴时期的背景。特别是在15世纪，欧洲社会发生了重大变化，欧洲的眼界无论在字面意义上还是比喻意义上都大为拓宽。四个关键事件或运动从根本上重新塑造了16、17世纪的人所生活的世界：人文主义的兴起、活字印刷术的发明、地理大发现和基督教改革。虽然不是严格意义上的科学发展，但这些变化为这一时期的思想家重新塑造了世界。

文艺复兴及其中世纪起源

“意大利文艺复兴”一词常使我们想起桑德罗·波提切利、皮耶罗·德拉·弗朗切斯卡、列奥纳多·达·芬奇、安吉利科等名人完成的艺术和建筑杰作。但文艺复兴远不只是美术的繁荣。文学、诗歌、科学、工程、民事、神学、医学及其他领域也得到了蓬勃发展。我们不应低估15世纪意大利文艺复兴时期的辉煌及其对历史和现代文化的重要性。但也应该记住，这并非公元5世纪罗马帝国陷落、古典文明灭亡之后欧洲文化的第一次重要繁荣。至少有两次更早的“复兴”或“重生”。

第一次是加洛林文艺复兴，发生在公元8世纪末查理曼的军事征服之后，它使中欧在公元9世纪的大部分时间里更为稳定。查理曼在亚琛的宫廷成了学问和文化的中心，为后来大学奠定基础的大教堂学校便是源于这一时期。公元800年，教皇利奥三世加冕查理曼为“罗马人的皇帝”，为加洛林改革定下了基调：试图回到古罗马的荣耀。建筑、造币、公共建筑甚至是书写风格都在有意模仿帝国时代的罗马人或至少是9世纪所想象的罗马人。不过这次繁荣很短命。

拉丁欧洲的第二次“重生”范围更广，持续时间更长。尽管强度逐步减弱，但其势头一直持续到意大利文艺复兴开始。这次“重生”即所谓的“12世纪文艺复兴”，科学、技术、神学、音乐、艺术、教育、建筑、法律和文学中的创造力喷薄而出。这一繁荣的起因尚有争议。一些学者指出，从11世纪开始的欧洲气候更为温暖宜人（被称为“中世纪暖期”），农业的进步使食物增多、经济繁荣，欧洲人口短时间内翻了一番甚至增至三倍。城市中心的兴起、更稳定的社会政治制度、更为充足的食物以及随之而来的从事思考和学术的更多时间，所有这些都有助于促成这次复兴。

觉醒的欧洲在穆斯林世界找到了丰富的思想资源。基督教欧洲开始在西班牙、西西里和黎凡特将伊斯兰教的边境往后推移时邂逅了阿拉伯的学术财富。穆斯林世界曾经继承了古希腊知识，将其译成了阿拉伯文，并用新的发现和观念数度丰富了它们。在天文学、物理学、医学、光学、炼金术、数学和工程领域，“伊斯兰聚居地”都远胜于拉丁西方。欧洲人坦然接受了这一事实，并立即致力于获取和吸收阿拉伯的学问。欧洲学者在12世纪开始了一场伟大的“翻译运动”。数十位翻译家（往往是修道士）长途跋涉来到阿拉伯世界特别是西班牙的图书馆，历经艰辛将数百部著作译成了拉丁文。具有特殊意义的是，他们选择翻译的文本几乎都是科学、数学、医学和哲学领域的文本。

拉丁中世纪只从古典世界继承了罗马人所拥有的那些文本。到了罗马帝国晚期，只有少数罗马学者能够阅读希腊文，因此罗马人所能传承的文本几乎只有对希腊学问的拉丁文释义、概述和普及。这就好比我们的后人只获得了新闻报纸对于现代科学的记述和普及，而几乎没有获得科学期刊或文本。于是，拉丁中世纪的学者尊崇古代伟大作者的名号，并拥有对其思想的描述，但几乎没有他们的著作。

12世纪的翻译家彻底改变了这一局面，他们翻译了阿拉伯学者的原创性著作和古希腊著作的阿拉伯文译本。大多数古希腊文本就这样披着阿拉伯的外衣传到了欧洲。阿拉伯文本贡献了盖伦的医学、欧几里得的几何学、托勒密的天文学以及我们今天所拥有的亚里士多德的几乎全部著作，更不用说阿拉伯学者在所有这些领域以及其他领域中更高阶的著作。1200年左右，这些激增的知识变成了大学中的课程，而大学也许是中世纪为科学和学术所留下的最为持久的遗产。亚里士多德的自然哲学著作构成了课程的核心，他的逻辑学著作促成了经院哲学，这是关于逻辑研究和争论的一套严格的形式化方法，可以运用于任何主题，大学研究正是以经院哲学为基础。

大学作为学术机构的重要性怎样强调都不为过。正如著名学者爱德华·格兰特所说，中世纪的大学"塑造了西欧的精神生活"。虽然大学中级别最高的是神学，但一个人如果不首先掌握当时的逻辑学、数学和自然哲学，就不可能成为一名神学家，因为这些论题经常被用在中世纪高级的基督教神学中。事实上，中世纪大多数伟大的自然哲学家都是神学博士，如大阿尔伯特（现在是自然科学家的主保圣人）、弗赖贝格的狄奥多里克、尼古拉·奥雷姆、朗根施泰因的亨利等等。他们全都在大学里学习和任教，并在那里找到了归宿。

14世纪的灾难阻碍了13世纪充满活力的文化生活。14世纪初，可能是由于"中世纪暖期"的结束，反复的粮食歉收和饥荒袭击了当时已人口过剩的欧洲。14世纪中叶，黑死病瘟疫突然席卷欧洲，一周之内很多人染病而死。就导致的生命损失或社会剧变而言，如今没有任何东西能像黑死病那样迅速、势不可当和具有破坏性。从1347年到1350年四年间，近一半欧洲人口命丧于此。意大利文艺复兴的最初迹象正是出现于这些动荡时期之前——诗人但丁（1265—1321）活跃于黑死病之前，比他年轻的作家薄伽丘（1313—1375）和彼特拉克（1304—1374）则活过这段时期幸存了下来。

人文主义

瘟疫盛期过后，在一两代人时间里发生的意大利文艺复兴为科学革命提供了第一个关键背景：**人文主义**的兴起。由于难以对人文主义作出简洁而严格的定义，我们最好谈及**复数的人文主义**（humanisms）——思想、文学、社会政治、艺术和科学上的一些彼此相关的潮流。人文主义者持有一种非常普遍的信念，认为自己生活在一个兼具现代性和新颖性的新时代，这个新时代应当结合古代人的成就加以衡量。他们部分是通过研究和仿效古代的希腊人和罗马人来寻求艺术与文学的复兴。据此，莱奥纳尔多·布鲁尼（1369—1444）和弗拉维奥·比翁多（1392—1463）等意大利文艺复兴时期的人文主义历史学家提出了我们所熟悉的三阶段历史分期（我们至今仍须努力从它的意涵中解放出来）。根据这种分期，第一个时代是古希腊罗马，第三个时代是现代，当然始于文艺复兴时期的作家们本人。根据人文主义者的说法，这两个高点之间是一个沉闷和停滞的“中间”时期，因此被称为“中”世纪。事实上，关于公元500年到1300年这一时期的所有名称无不充斥着意大利人文主义者对它的鄙视，就此而言，文艺复兴时期最为持久的发明也许就是“中世纪”这一概念。鉴于人文主义者的直接背景就是对饥荒和瘟疫之年的切近记忆，1400年左右意大利的重新繁荣必定像是一个“新时代”的黎明。

模仿被视为最真诚的奉承，人文主义者通过模仿罗马风格来表达他们对古代的仰慕。以前也曾有过回到古代的尝试，特别是在600年前的加洛林文艺复兴时期。罗马的壮观的确给人类的记忆留下了深刻的印象。人文主义者渴望更多地了解那个过去的时代，这表现于对久已遗失的古典文本的寻求。早期的人文主义者波吉奥·布拉乔利尼（1380—1459）利用具有革新意识的康斯坦茨会议（1414—1418，他担任教皇秘书）的休会期，遍寻附近的修道院图书馆，以寻觅幸存下来的古典文献。他不仅发现了昆体良论修辞的著作以及此前不为人所知的西塞罗演说，而且——对于科学史更重要的是——也发现了卢克莱修介绍古代原子论思想的《物性论》、马尼留斯的天文学著作、维特鲁威的建筑和工程著作以及弗龙蒂努斯论述水道和水力学的著作。数百年来，这些作品经由中世纪修道士的抄写——也许只剩下了某个孤本——而在修道院的图书馆中一代代保存下来。

人文主义者对罗马学问的重新恢复伴随着希腊语研究的复兴。拉丁西方在一千年的时间里几乎完全不通晓希腊语。希腊语复兴的背景是希腊外交官和教士代表团于1400年左右来到意大利。他们的使命是获取援助以抵制土耳其人的威胁，使1054年以来分裂的东西方教会重新联合起来。克利索罗拉斯（约1355—1415）是最早来到意大利的外交官之一，但他转而在那里讲授希腊语，许多著名的人文主义者都成了他的学生。意大利人对希腊文本的渴望被激起，他们继而前往君士坦丁堡搜寻手稿。瓜里诺·达·维罗纳（1374—1460）带回了数箱手稿，其中包括斯特拉波的《地理学》，随后被他译成了拉丁文。据说曾有一箱手稿在运输过程中丢失，瓜里诺达因此过于悲伤而一夜白头。参加15世纪30年代佛罗伦萨会议的希腊代表团包括两位著名的希腊学者。一位是后来做了红衣主教的约翰内斯·贝萨里翁（1403—1472），他将自己收集的近一千份希腊手稿赠予了威尼斯城。另一位是乔治·盖弥斯托斯（约1355—约1453），一般被称为普勒托，这位性格乖张的学者后来倡导回归古希腊多神教。普勒托在佛罗伦萨教希腊语，并使西方注意到了柏拉图和柏拉图主义者的著作。他的教导促使执政的大公科西莫一世·德·美第奇在佛罗伦萨创建了一个柏拉图学院。学院的第一位领导者菲奇诺（1433—1499）翻译了柏拉图的著作以及后来几位柏拉图主义者的文本，其中大部分作品当时还不为西欧读者所知。

于是，和12世纪一样，15世纪也重新发现了大量古代文本，其中许多是关于科学技术主题的。但人文主义者的区别性特征与其说是热爱文本，不如说是热爱**纯粹而准确的**文本。他们称大学中使用的亚里士多德和盖伦的文本是不纯的——充满了野蛮、“阿拉伯特征”（Arabisms）、添加和错误。他们将经院哲学斥为贫瘠的、野蛮的和不雅的。他们认为，大学（尤其是北方的大学，意大利的情况要好一些）是停滞的“中间”时代的遗迹，斥责大学学者在写一种退化的拉丁文，缺乏优雅的气质。因此，人文主义的一个重要特征是在大学以外建立了新的学术共同体。

一个现代误解是，人文主义者由于某种原因是世俗主义的、非宗教的甚至是反宗教的。一些人文主义者固然会批评教会的恶习，蔑视经院神学，但他们决不反对基督教或宗教。事实上，许多人倡导的教会改革与他们期待的语言改革相平行——通过回到古代、回到公元后最初几个世纪的教会来实现。许多人文主义者都担任圣职，在教会机构任职，或者享受教士俸禄，天主教的等级结构支持了人文主义。文艺复兴时期的多位教皇都是热情的人文主义者，尤其是尼古拉五世、西克斯图斯四世和庇护二世，他们的红衣主教和宫廷也是如此，都鼓励人文主义者。现代的错误缘自将其与所谓的**世俗人文主义**相混淆，这是一项20世纪的发明，在近代早期并无与之对应的概念。

文艺复兴时期的人文主义对科学技术史的影响正负兼有。从积极的一面来讲，人文主义者得到了数百个新的重要文本，使考据学达到了新的水平。对柏拉图的重新引入（特别是由于他采用的毕达哥拉斯数学）提升了数学的地位，并且提供了一种与大学中受到青睐的亚里士多德主义不同的哲学。为了符合古人的标准，整个意大利的工程和建设项目均以古代工程师阿基米德、希罗、维特鲁威和弗龙蒂努斯为典范。其消极一面是，对古代的奉承可能走得太远，以致将罗马帝国灭亡之后的一切事物都斥为野蛮。于是，欧洲开始失去对阿拉伯和中世纪成就的尊重和认识，而阿拉伯和中世纪在科学、数学和工程领域的成就——毫无疑问——大大胜过了古代世界。

印刷术的发明

1450年左右活字印刷术的发明极大地满足了人文主义者对文本的兴趣。这一发明（或至少是其成功推广）要归功于约翰内斯·古腾堡（约1398—1468），他原本是美因茨的一名金匠。活字印刷术的关键是铸造带有突出字母的金属活字。这些活字可以组装成完整的文本页面，在其表面涂上一种油墨，按压在纸上，一次便可印刷一整页（或一组页面）。印刷多份之后，可将活字页面拆卸开，很容易把字母重新排列成下一组页面。此前，书籍必须手工抄写，造成产量低下且价格高昂。中世纪晚期大学的发展和大众读写能力的增长使书籍供不应求，人们对生产书籍的速度有了更多需求，一批制书企业在修道院和大学的传统缮写室之外应运而生。产量的增加导致了越来越多的抄写错误，人文主义者对此深感痛惜。印刷术确保了图书生产的效率和质量，尽管花费在造纸、排版和印刷上的劳动使得书籍仍然十分昂贵。（1455年印出的古腾堡《圣经》价格为30弗罗林，比一个熟练工匠一年的工资还要多。）

从抄写到印刷的过渡并非一蹴而就。手稿仍然和书籍并存，虽然手稿的使用越来越多地局限于私人的、稀有的或特许的材料。印刷字体模仿手稿书写；在北欧，这意味着用哥特字体印刷书籍，但意大利（特别是威尼斯）很快就成了印刷业的中心。意大利印刷工泰奥巴尔多·曼努奇［其拉丁化的人文主义名字马努提乌斯（1449—1515）更为人所知］采用了意大利人文主义者发明的更为干净清晰的字母形状（他们认为模仿了罗马人的书写方式），由此创造的字体不仅取代了旧字体，而且也是今天使用的大多数字体的基础；因此，我们优雅的斜体仍然被称为“意大利体”（Italic）。

印刷机如雨后春笋般地出现于整个欧洲。到了1500年，大约有1000部印刷机在运转，已有三四万种图书被印制出来，总数约有1000万册之多。这些印刷制品在整个16、17世纪有增无减。书籍变得越来越便宜（往往质量有所下降），更容易为不太富裕的购买者所获得。印刷使得通过相互攻讦、时事通讯、小册子、期刊和其他生命短促的纸质媒介进行的交流得以更快地进行。虽然这些纸质媒介生产出来以后大都很快就消亡了（如上周的报纸），但它们在近代早期是非常普遍的。就这样，印刷机创造出一个史无前例的印刷文字的新世界，一个读写文化的新世界。

印刷的一个容易被忽视的特点是它能够复制**图像和图表**。在手稿传统中，插图是一个问题，因为准确绘图的能力取决于抄写者的技法，而且经常依赖于他对文本的理解。因此，无论是解剖图、动植物插图、地图、海图、数学图解还是技术图解，每一次复制都意味着质量的下降。一些抄写员径直忽略了困难的图形。印刷意味着作者能够监督生产原版的木刻或雕刻，然后便可以轻松可靠地生产完全相同的副本。在这种情况下，作者更愿意并且能够把图像包括在他们的文本中，从而促成了科学插图的第一次发展。

航海大发现

一张图片胜过千言万语，事实证明，绘制插图的能力特别重要，因为新奇的报告和事物很快就会涌入欧洲。这些信息来自欧洲人直接接触的新土地。第一个来源是亚洲和撒哈拉以南的非洲地区。由于葡萄牙人试图开辟与印度的贸易航线，以便绕过控制了陆路和地中海航线的中间商（主要是威尼斯人和阿拉伯人），遂使欧洲人接触到了这些地区。15世纪初，葡萄牙王子、航海家亨利（1394—1460）开始派远征队沿西非海岸探险，与撒哈拉以南非洲地区的商人建立了直接联系。葡萄牙水手进一步南下，最终于1488年绕过好望角，其最高潮是达·伽马于1497至1498年成功远航至印度进行贸易。葡萄牙人沿途建立了贸易前哨，其中许多地区直到20世纪中叶仍为葡萄牙所拥有，他们最终将其常规航线延伸到中国，将香料、宝石、黄金、瓷器等奢侈品运回欧洲，还带回了关于遥远国度、奇异生物和未知民族的报道。

欧洲视野的这种拓宽并非在文艺复兴时期遽然开始。中世纪为文艺复兴时代的航海奠定了基础。事实上，向东的航行早在13世纪就已出现，却因14世纪亚洲的政治动荡而被迫中断，到了15世纪又被恢复。中世纪的旅行者往往是13世纪两个新修会——多明我会和方济各会——的成员，他们开始到遥远的地方传教和从事外交活动，这种使命我们直到现在才有所认识。他们在亚洲建立了宗教场所，从波斯和印度一路推进到北京，并将相关信息传回欧洲，从而激励了后来的贸易航行。这些中世纪旅行使人们意识到欧洲之外还有一个更为广袤的世界有待探索。

当葡萄牙人正在向东开辟朝向亚洲的海上航线时，哥伦布却把目光投向了相反方向。他确信，地球周长大约要比在欧洲广为人知的相当准确的古代估计值短三分之一，因此认为自己向西航行能够到达东亚。这种错误的印象部分是由于公元2世纪的地理学家和天文学家托勒密。人文主义者们刚刚重新找出了他的《地理学》，其中把地球的尺寸说得异常之小，大大高估了亚洲向东的范围。哥伦布的资助者持怀疑态度，他们认识到西行路线要更长，如果没有中间的地方提供新的补给，船员就会饿死。（**没有人**认为哥伦布会"航行到地球边缘掉下去"，因为早在哥伦布之前1500多年，地球的球形观念已在欧洲牢固确立。说哥伦布之前的人都认为地球是平的，这是19世纪的发明。中世纪的人会对这种想法捧腹大笑！）因此，当1492年哥伦布的船只突然发现加勒比地区的陆地时，他自认为到了亚洲，而不是发现了一个新大陆。

无论哥伦布后来是否承认了自己的错误，其他人反正很快认识到了，于是急忙赶往这个新世界。在新发明的印刷机的帮助下，新世界的消息迅速传开。1507年，一位德国制图师根据意大利探险家亚美利哥·韦斯普奇的名字给这块新大陆命名为亚美利加。由于这些地图以及韦斯普奇随之发表的关于南美的描述，这个名字流传开来。1508年，西班牙国王费迪南多二世为韦斯普奇设立了新世界首席航海家一职。这一新职位所属的商局（Casa de Contratación）成立于1503年，不仅是为了给带回西班牙的货物征税，而且也是为了对返回的旅行者所带来的各种信息加以收集和分类，训练领航员和航海家，以及用从每一位返回的船长那里新收集到的信息不断更新原版地图。各种知识和技术诀窍汇集到塞维利亚，帮助西班牙建立了历史上第一个"日不落"帝国。

面对西班牙和葡萄牙正在积累的领地和财富，其他国家也不甘心袖手旁观，遂纷纷加入竞争行列，尽管他们落后于古伊比利亚人一个世纪或更长时间。因此在一百年的时间里，几乎所有关于新世界的报道和样本都是经由西班牙和葡萄牙来到欧洲的，它们改变了欧洲人的动植物知识和地理学知识。很难想象从新世界大量涌入欧洲的材料有多少。新的植物、动物、矿物、药品以及关于新的民族、语言、思想、观察和现象的报道使旧世界目不暇接，难以消化。这是真正的"信息过剩"，它要求改变关于自然界的想法，用新方法对知识加以组织。由于发现了新的奇异生物，传统的动植物分类系统不再适用。由于发现人类的居住地几乎无处不在，那种古代观念遭到了驳斥，即世界被分为五个气候区，包括两个温带和三个因为过热或过冷而不适宜居住的区域。开发美洲和亚洲巨大的经济潜力需要新的科学技术。地理数据和航线记录催生了新的绘图技术，而在欧洲与新国度之间安全可靠地通航则需要改进导航、造船和军备。

基督教改革

环游世界使欧洲人看到了各种不同的宗教观点，而宗教观点在欧洲本土也开始变得多样化。1517年标志着基督教内部开始出现一种深刻的、往往伴随着暴力的持续分裂。那一年，奥古斯丁会的神职人员和神学教授马丁·路德（1483—1546）在维滕贝格大学城提出了著名的《九十五条论纲》。这些论纲或命题以经院论辩主题的格式写成，集中批判了当地出售赎罪券的不当做法，这种做法在神学上是站不住脚的。虽然关于仪式和教义问题的类似争论在中世纪大学的论辩文化中很常见，但路德的抗议超出了神学学术争论的通常界限，迅速演变成一场超出马丁·路德控制的、有广泛基础的政治社会运动。路德的主张最初很温和，但逐渐变得越来越大胆和有对抗性，从地方做法的一些小问题升级为严重的教义问题。这些主张经由印刷机迅速传播开来，因与地方民族主义的联系而加深，并且受到了德国统治者的唆使，他们认为脱离罗马对其政治利益有利。就这样，一次地方性的抗议（protestation）出乎预料地发展成了新教（Protestantism）。新教几乎立刻分裂成了若干相互争论不休的派别。除了天主教与路德教的争论，很快又出现了路德教与加尔文教的争论，然后是加尔文教内部的争论，等等。所谓的“宗教战争”——激励它的往往更多是政治和王朝的操纵，而不是教义问题——在接下来的一个半世纪里震撼着欧洲，特别是德国、法国和英国。

路德本人并非人文主义者，尽管他的一些理念，如强调对《圣经》的字面理解而不是天主教徒所青睐的隐喻读法，与人文主义者对文本的强调有相似之处。但比这些相似之处更重要的是他怀疑古典的（“异教的”）文献和思想，并且希望把《圣经》中那些不同于其个人观念的各卷（如《雅各书》）删除。然而，比他有学识得多的梅兰希顿（1497—1560）却完全不是这样。

“梅兰希顿”这个名字证明了他的人文主义，它是从原本粗野的德国“黑土地”（Schwartzerd）翻译成的古典希腊文。提出这种“自我古典化”的是他伟大的伯父、德国最引人注目的人文主义者罗伊希林。紧随路德对大学经院哲学的拒斥，梅兰希顿（作为一个同样不喜欢经院哲学的人文主义者）调整了从天主教皈依路德教的德国大学——特别是路德本人所在的维滕贝格大学——的课程设置和教学。他设计的新课程使他赢得了“日耳曼之师尊”的头衔。其方法并非驱逐亚里士多德，而是——以真正人文主义的方式——消除中世纪向亚里士多德所作的“增添”以及使用更佳版本的希腊哲学家著作。新兴的新教大学不得不重新开始，淡化业已确立的方法，从而能把在旧体制中无法立足的新的主题和研究方法包括进来。

天主教内部的改革运动也在进行。在15世纪，宗教会议解决了一些问题，虽然不是很成功。更引人注目的是特伦托会议（1545—1563），这次大公会议通过处理腐败问题、澄清教义、规范仪式、集中纪律监督等做法，对新教作出了回应。直到第二届梵蒂冈大公会议（1962—1965），特伦托会议一直是中世纪之后最重要的教廷会议，它拉开了天主教改革或者说“反宗教改革”的序幕。其措施包括改进教士的教育（这一改革是许多人文主义者所提倡的）以及加强对于发表作品中正统学说的监督。一个新组建的教士团体——耶稣会最积极地参与到了特伦托会议所倡导的改革之中。1540年，圣依纳爵·罗耀拉在教皇的授权下建立了耶稣会，耶稣会士们尤其致力于教育和学术，在科学、数学和技术等领域做出了重要贡献。

除了宣扬新教徒回归天主教，耶稣会士更广泛的影响在于他们在创会最初几年所建立的数百所学校和学院。耶稣会的教育基于一种新颖的教学和课程风格，它坚持了亚里士多德方法的重要性，但重新强调了数学（到1700年，耶稣会士占据着欧洲一半以上的数学教授职位）和科学。科学革命的一些新科学思想往往是在耶稣会学校最先讲授的，许多孕育这些观念的思想家便是在这些学校培养出来的。耶稣会士沿着新开辟的贸易路线前往世界各地，高姿态地进入了中国、印度和美洲（当然包括开办学校），建立了第一个全球性的通信网络。该网络把一切事物都带回了罗马，无论是生物标本、天文观测和文化制品，还是关于本土知识和风俗的广泛报道。耶稣会对于研究科学和数学的态度表达了它的座右铭："在万事万物中找到神。"虽然耶稣会士强调这种激励，但这并非他们所独有，而是几乎整个科学革命的基础。

16世纪的新世界

16世纪的欧洲人居住在一个迅速变化的新世界。和我们快节奏的今天一样，许多人认为这种状况是焦虑的来源，而另一些人则看到了一个充满机遇和可能性的世界。欧洲的视野在各种意义上得以拓宽。欧洲人重新发现了他们自己的过去，遇到了一个更广的物理世界和人类世界，创造了新的研究方法，对旧观念作了新的诠释。事实上，用一个喧哗骚动、储备丰富的市场来形容他们的世界再恰当不过。纷杂刺耳的声音大大促进了各种思想、货物和机遇。人们摩肩接踵地对各种商品进行检验、购买、拒绝、赞美、批评或只是触碰。几乎所有东西都供人竞购。无论我们认为"科学革命"是某种全新的东西，还是经过14世纪的不幸中断之后对中世纪晚期思想发酵的恢复，毫无疑问的是，16、17世纪有学识的居民都认为自己处于一个充满变化与新奇的时代。这是一个激动人心的时代，一个新世界的时代。

第二章

关联的世界

近代早期思想家看到的世界是真正希腊意义上的**宇宙**（cosmos），即一个秩序井然、恰当安排的整体。在他们眼中，物理宇宙的各个组成部分彼此密切交织在一起，并且与人和神紧密相关。他们的世界织成了一张关联和相互依存的复杂网络，它的每一个角落都充满了目的、密布着意义。因此对他们而言，研究世界不仅意味着揭示其内容事实并加以分类，而且意味着揭示其隐秘设计和无声的寓意。这种看法与现代科学家形成了鲜明的对比，日益专业化使现代科学家只关注那些狭窄的研究主题和孤立的对象，其方法强调切割而不是综合，其态度主动排除了意义和目的问题。现代研究方法成功地揭示了关于物理世界的大量知识，但也造就了一个脱节的、支离破碎的世界，使人类感到疏离和孤立于宇宙。几乎所有近代早期自然哲学家都持有一种更为广泛和无所不包的世界观，他们的动机、问题和做法正是源于这种视野。因此，要想理解他们研究世界的动机和方法，就必须理解他们的世界观。

一个内在紧密关联的目的论世界的观念有许多来源，但最重要的来源是柏拉图和亚里士多德这两位不可回避的古代巨人以及基督教神学。从柏拉图特别是所谓的晚期柏拉图主义者或新柏拉图主义者——基督教时代最初几个世纪在希腊化的埃及积极发展柏拉图思想的哲学家——那里产生了“自然阶梯”（*scala naturae*）的思想。根据这种构想，万事万物都在一个连续的层次结构中拥有特殊的位置。其顶端是太一，完全超越的永恒的神，其他一切事物的存在都源于此。太一流溢出创造性的力量，使其他一切事物得以产生。这种力量越是从其来源流溢出来，它所创造的东西就越低和越不像太一。其底部是惰性的、毫无生气的质料。其间的等级按升序排列依次是植物和动物的生命，然后是人类，然后是精神性的存在，如精灵（*daimons*）和较小的神。一些新柏拉图主义者的目标仿佛是爬上阶梯，变得更具精神性和较少物质性，将人的灵魂——我们最崇高的部分——从堕入物质所导致的盲目性中摆脱出来，经由精神性存在的层次朝着太一上升。这种古代晚期观念和基督教教义相互影响，正如公元5世纪的柏拉图主义基督徒伪狄奥尼修斯所指出的，用不同等级的天使取代异教的精灵和较小的神，用基督教的上帝取代太一，很容易使这种观念符合正统基督教信仰。由于这种基督教化，自然阶梯的观念在整个拉丁中世纪都广为人知，尽管它所基于的古代柏拉图主义文本已经散佚了许多个世纪。

文艺复兴时期的人文主义者重新发现了这些柏拉图主义文本，并由菲奇诺译成了拉丁文。菲奇诺也获得、翻译和出版了一批以“三重伟大的赫尔墨斯”（Hermes Trismegestus）命名的文本，其作者被假想成一位与摩西同时代的古埃及圣贤。大约从公元前3世纪到公元7世纪，大量不同版本的《赫尔墨斯文集》产生出来，菲奇诺所获得的只是其中一小部分。这些文本虽然最初被认为要古老得多，但菲奇诺的《赫尔墨斯文集》实际上可能写于公元2世纪和3世纪。其重要性在于它的新柏拉图主义特征，强调了人类的力量，人类在关联的阶梯世界中的位置，以及人类沿阶梯向上攀升的能力。许多文艺复兴时期的读者在《赫尔墨斯文集》中找到了他们认为的基督教的预示，三重伟大的赫尔墨斯因此成了一位异教的先知，锡耶纳大教堂中描绘的众先知中就有他。

在对世界的**阶梯**设想中，任何被造物都有一个位置，都与它之上或之下紧挨着的被造物相关联，因此沿着所谓"伟大的存在之链"从最低层次到最高层次有一个渐进的、无间隙的连续上升。一个相关的概念——存在于柏拉图论述宇宙起源的《蒂迈欧篇》中，这是拉丁中世纪所知的柏拉图的唯一作品——是**大宇宙**和**小宇宙**。这两个希腊词分别意味着"大有序世界"和"小有序世界"。大宇宙是宇宙的身体，亦即恒星和行星的天文学世界，而小宇宙则是人的身体。其基本思想是，这两个世界的构造基于类似的原则，因此彼此之间存在着密切的关系。公元8世纪的一部被称为《翠玉录》的阿拉伯作品是对《赫尔墨斯文集》的一项晚期贡献，它将这一观点简洁地概括为近代早期欧洲众所周知的一则格言："上行下效。"对于柏拉图而言，将人的小宇宙与行星的大宇宙联系起来有一种实际的道德意义——我们应把天界有序而合理的运作视为指导，以一种有序、合理的方式来支配自己。对于近代早期的欧洲人而言，小宇宙—大宇宙的联系首先有一种医学意义——它是医学占星术的基础。各个行星对特殊的人体器官会产生特殊影响，从而影响人体的功能（见第五章）。

对于内在关联和目的论的世界观的第二项主要贡献，源自亚里士多德关于如何获得知识的想法。根据亚里士多德的说法，关于事物的严格知识是"因果知识"。我们需要对这个词作出解释。亚里士多德认为，要想认识一个事物，需要确定其四种"原因"或者说存在的理由。第一个原因是**动力因**，它描述了制成该物体的是什么或谁。**质料因**描述了该物体是由什么构成的。**形式**

因说明了是什么物理特征使物体是其所是，亦即其性质的集合。对于亚里士多德主义者来说，最重要的原因是目的因，这也是现代人最难理解的原因。**目的因**告诉我们事物是为了什么目的，即其现有的目标是什么，在亚里士多德看来，任何事物都有一个目标或目的。我们可以用阿基里斯的雕像来说明这些“原因”。这尊雕像的动力因是雕塑家，其质料因是大理石，其形式因是阿基里斯的美丽身体，其目的因是为了纪念阿基里斯。每一种原因可能不止一个（例如，雕像还可能有作为装饰，或者在一些雅典式房屋内作为衣帽架的目的因）。

关键的一点是，亚里士多德意义上的知识，特别是关于动力因和目的因的知识，**在物体相对于其他物体的关系背景下**对物体作出了定义。认识一个事物意味着找到它在与其他事物，尤其是产生它并且利用它的事物的关系网络中的位置。在欧洲的基督教背景下，目的因与神的设计和神意的观念非常协调。自然中的目的因是上帝创世计划的一部分，第一动力因将该计划植入了受造物并对其进行编码。

近代早期的作者以许多不同方式表达了他们对一个关联世界的理解。因化学工作而著名的英国自然哲学家玻意耳（1627—1691，学习化学的学生仍然要学玻意耳定律，即气体的体积与所施加的压力成反比）指出，世界就像是一部“精心构思的小说”。这里玻意耳暗指他非常喜欢读的那个时代的许多法国小说。这些小说的长度往往超过了2000页，并有许多令人眼花缭乱的主要角色，其复杂的故事情节不断以令人惊讶的方式收敛和发散，字里行间都在透露谁偷偷爱上了谁，谁是真正失散已久的兄弟、子女或无论什么事物。对玻意耳而言，造物主便是最终的小说作家，科学研究者则是那些读者，试图弄清楚造物主在世界中书

图1 基歇尔，《磁石，或者论励磁的方法》（罗马，1641）标题页版画，表明各个知识分支以及上帝、人和自然之间的内在关联。

写的所有关系和错综复杂的故事情节。

极为博学的耶稣会士基歇尔（1601/1602—1680）在罗马维护着一座奇迹博物馆，他是耶稣会士就自然哲学进行通信的一个中心。在他关于磁学的一本百科全书式的著作中，一幅优雅的巴洛克风格的卷首插图（图1）描绘了这种内在关联的世界。

该图显示了一系列圆形印章，每一个印章上都带有某个知识分支的名称：物理学、诗学、天文学、医学、音乐、光学、地理学等等，神学则在最顶端。一个链条将这些印章连接在一起，表达了所有知识分支的内在统一性。对于近代早期的人而言，并没有什么严格的壁垒使科学、人文学科和神学彼此隔绝，它们形成了探索和理解世界的环环相扣的方法。在基歇尔的图像中，有链条将这些知识分支与三个更大的印章连在一起，后者代表自然世界的三个主要部分：天界（比月球更远的一切事物）、月下世界（地球及其大气层）和小宇宙（人）。世界的这三个部分也同样连接在一起，表明它们之间存在着无可避免的相互依存关系。整个图像的中心处是分别与三个世界直接接触的原型世界（mundus archetypus），即上帝的心灵，它不仅创造了世间万物，而且包含着宇宙中一切可能事物的模型或原型。基歇尔以一句拉丁文格言完成了这幅图像："由隐秘之结关联起来的万物平静地安歇着。"

这种对各个学科之间以及宇宙各个方面之间关联性的感受是**自然哲学**的典型特征。自然哲学是近代早期自然研究者所从事的学科，它与我们今天所熟悉的**科学**密切相关，但在范围和意图上更加广泛。中世纪或科学革命时期的自然哲学家和现代科学家一样研究自然界，但这种研究是在包括神学和形而上学在内的更广泛的视野中进行的。神、人、自然这三个组成部分从未彼此隔绝。到了19世纪，自然哲学观念逐渐让位于更加专业化的狭窄“科学”视角，“科学家”一词正是在此时期被创造出来。如果不时刻牢记自然哲学的独特性，就不可能正确理解或欣赏近代早期自然哲学家的工作和动机。他们的问题和目标并不一定是我们的问题和目标，即使研究的是同样的自然对象。因此，我们撰写科学史时绝不能把那些科学上的“第一次”从历史背景中抽离出来，而只能通过历史人物的眼睛和心灵去审视它们。

自然“魔法”

在16、17世纪得到广泛认同的这种“整体宇宙”观是各种努力和事业的基础，即使不同思想家认为世界中的内在关联对其工作有不同程度的重要性。在自然哲学中，与这种世界观联系最紧密的一个方面是自然魔法（*magia naturalis*）。把这个拉丁词直接译成英语的“natural magic”是一种误导，因为“magic”一词很容易使现代读者想起特殊打扮的人从帽子里掏出兔子，或者头戴尖顶帽子、身披黑色长袍的皮肤干皱的人在大锅前喃喃自语，或者更加亲切地想起哈利·波特和霍格沃茨魔法学校。而近代早期的自然魔法则非常不同，它是科学史的重要组成部分。

对于现代人来说，*magia*（魔法）最好的译法也许是“控制”（mastery）。践行魔法者被称为魔法师（magus），其目标是学习和控制内嵌于世界中的各种关联，以便出于实际目的对其进行操纵。再看看基歇尔的卷首插图。在左上角，自然魔法被列为一个知识分支，介于算术和医学之间。基歇尔用向日葵转向每天在天空中行进的太阳来象征它。（有几种植物显示出这种被称为**向日性**的行为。）为什么向日葵总是转向太阳，而大多数植物却不这样？显然，太阳与向日葵之间必定存在着某种特殊的关联。

向日葵能够跟随太阳，这种能力为世界中隐秘的关联和力量提供了一个绝好例子，魔法师力图确认和控制的正是这些关联和力量。

中世纪的亚里士多德主义者将事物的性质分为两组。第一组是**明显性质**，即任何有感觉器官的人都能觉察的性质。热、冷、湿、干是首要性质。其他性质包括光滑、粗糙、黄、白、苦、咸、响亮、芳香等等所有激活感官的东西。毕竟，亚里士多德主义从根本上讲是一种常识性的与世界打交道的方式。亚里士多德主义者用这些明显性质来解释一个事物对另一个事物的作用，例如冰凉饮料之所以能够退烧是因为冷能够抵消热。但有些物体起作用的方式比较怪异，显示出一些无法解释的性质。这些物体被认为具有我们无法用感官觉察到的**隐秘性质**（*qualitates occultae*，常被误导地译为"神秘性质"）。这些性质常以非常特定的方式起作用，暗示特定事物与其作用对象之间存在着一种特殊的无形关联。中世纪的自然哲学家列出了一系列此类现象。一个典型的例子是磁铁。关于磁石（一种天然的磁性矿物），我们感觉不到任何东西能够解释它吸引铁的神秘能力。太阳与向日葵之间的表观吸引力、罗盘针指向北极星、鸦片的催眠作用、月亮对潮汐的影响以及其他许多事物也是如此。自然魔法便是要努力找出事物的这些隐秘性质及其效应并加以利用。

如何在自然中发现这些关联、这些"隐秘的结"呢？一种方法是近距离地观察这个世界。每个人都会同意，认真观察是科学研究的一个关键出发点；寻求自然魔法促进了这样的观察。同样重要的一种方法是发掘早期自然观察者的记录——古往今来各种文本记录中或平凡或怪异的叙述和观察。因此，许多魔法都要基于对文本进行人文主义式的认真解读，通过搜集早期作者的说法而建立起复杂的网络。鉴于自然的无限多样性，雄心勃勃的魔法师的任务宏大得令人难以置信——几乎是为所有事物的属性进行编目。是否可能存在一种捷径？一些自然哲学家相信自然中包含着若干线索来引导魔法师，这些线索也许由一个仁慈的上帝植入自然，希望我们明白他的创造并从中获益。**征象学说**（doctrine of signatures）声称，一些自然物被“签署”（signed）了显示其隐秘性质的迹象。这往往意味着两个有关联的物体看起来有些相似，或者有一些类似的特征；例如，向日葵不仅追随太阳，而且其颜色和形状也**类似于**太阳。植物的各个部分好似人体的各个部分；外壳中的核桃仁看起来很像颅骨中的大脑。这是否暗示着核桃补脑呢？魔法师固然要试验这些东西，但观察以及征象的观念为研究、解释和利用自然界提供了一个有益的出发点。

征象学说只是代表着近代早期无处不在的一种更广泛的类比思维模式的一个方面。现代人往往会把这些相似之处看成仅仅是巧合或偶然，或者看成“诗意的”而不是物理的，但近代早期的许多人看待事物的方式却完全不同。他们**预料**世界的各个部分之间存在着类比关联，对他们来说，发现自然之中存在着一种类比或对称就意味着事物之间存在着一种实际关联。两个自然物之间的每一种类比绝非人类想象力的产物，而是标示出了创世蓝图中的又一条线，是上帝植入宇宙的一种隐秘关联的可见迹象。因此，类比论证所具有的特殊力量和明证性超出了我们今天的习惯看法。这种联系的确实性乃是基于一种不可动摇的信念，即相信宇宙不是随机或偶然的，而是充满着意义和目的，它由神的智慧和意志以多种方式引导，最终是为了人类的利益。

这种确定性以及随之而来的对类比推理的运用并非为那些对自然魔法感兴趣的人所独有，而是属于这一时期几乎**每一位**严肃的思想家。

运用直接观察、类比、权威文本和征象，近代早期思想家搜集了大量他们认为存在关联的事物。例如，还有什么可能关乎太阳与向日葵的关联？太阳是大宇宙的热源和生命之源，它在小宇宙中的对应部分必定是心脏。（再看看基歇尔的卷首插图——在代表小宇宙的人体的心脏位置有一个小太阳。）太阳是最高贵的天体，光辉夺目，呈现明亮的黄色，类似于矿藏中的黄金，并进而类似于所有黄色或金色的东西。在动物领域，太阳使公鸡打鸣，表明两者之间有一种特殊关联。狮子的黄褐色、百兽之王的地位以及类似于太阳的头部（狮子头上的鬃毛宛如太阳射线）似乎也与太阳有关联。同样，狮子的勇敢又对应于心脏。太阳、向日葵、心脏、黄金、黄色、公鸡和狮子都有某些共同特性，因而存在着实际却又隐秘的关联。在自然魔法的倡导者看来，可以把这些类比关联变成可利用的操作性关联。最实际的应用是把黄金或向日葵用作治疗心脏的药物——但我们将会看到，事情可能变得更富戏剧性。

究竟是什么东西把束缚在这些相似性网络之中的物体关联起来，人们对此有不同看法。但这些物体通常被认为是通过“共感”（sympathy）起作用，其字面意思是“一起遭受或一起接受作用”。考虑两张调好音的鲁特琴，分别位于房间两侧，拨动其中一张琴的弦，则另一张琴相应的弦将立即开始振动并发出嗡嗡声，与拨动第一张琴发出的声音相呼应。今天，我们仍然称这种现象为**共**振（sympathetic vibration）。对于近代早期思想家来说，这种现象体现了空间上分离的彼此“合调”的两种东西之间看不见的关联。有些人认为，空间上分离的东西之间要想传递作用必须通过介质；亚里士多德指出，如果没有一种居间的介质传递效果，一个物体就不可能作用于另一个有空间距离的物体。例如就琴弦而言，我们知道居间的空气传递着两个乐器之间的振动。对于其他共感作用，该介质可能是所谓的“世界精气”（*spiritus mundi*）——一种普遍的、渗透一切的、无形的或准有形的东西，通过把影响从一个物体传到另一个物体，它能使相距遥远的物体实质上彼此接触。这种“精气”并不是某种具有感知能力的超自然的东西，而是大宇宙中与小宇宙的“生命精气”等价的东西，“生命精气”是我们身体之中一种精细的东西，当我们的理智意识到有一辆两吨重的卡车正在加速向我们驶来时，“生命精气”经由神经将“快跑”的命令传到我们的脚。“世界精气”也类似地把“信号”从太阳传到向日葵，或者从月亮传到海水。大宇宙和小宇宙再次彼此映射，两者都包含着传递信号的精气。顺便提及，这种相似性也意味着大宇宙本身拥有某种灵魂——柏拉图在《蒂迈欧篇》中断言了这一点，现代人尤其难以理解——下一章我们会回到这一点。

从厨房到书房的实践“控制”

关于关联世界的自然魔法理论令人印象深刻，堪称优雅和美妙，但自然魔法的关键特征在于实际应用。近代早期魔法的实践部分既有平常的也有崇高的，前者往往没有任何理论基础。德拉·波塔（1535—1615）的《自然魔法》一书便是一个很好的例子。德拉·波塔因为在那不勒斯建立了最早的科学社团——秘密学院——和身为猞猁学院的一员而闻名，猞猁学院是17世纪初的科学社团，伽利略即为其中一员。德拉·波塔《自然魔法》的

第一章概括了一个内在关联的世界的原理，并指出魔法为何“是对整个自然进程的考察”以及“自然哲学的实践部分”。他建议读者“大力对事物进行探究；既要积极认真研究，又必须耐心等待。……必须不遗余力地做事，因为自然的奥秘不可能透露给懒惰的闲人”。德拉·波塔的书的其余部分所揭示的实际自然奥秘的确包括对磁学和光学的考察，但该书的大部分内容却是各种秘方和诀窍，从制作人造宝石和烟花爆竹，到动植物育种，再到关于制作香料、烤肉、水果保鲜等等的家用建议，其中没有任何东西利用了关于世界的理论观念。德拉·波塔的书符合一种“秘密之书”的传统，这种传统在整个16、17世纪变得越来越流行，而这些“秘密之书”中有一部分直到19世纪还被重印。许多此类书籍都是先来阐述关于宇宙的宏大而崇高的观念，但主要内容是家庭管理或家庭手工业的诀窍，并不包含或几乎不包含关于世界本性的内容。

菲奇诺（1433—1499）处于该阶梯的崇高一端，他将世界关联性的实际应用体现在生活方式和仪式中。菲奇诺经常抱怨自己的忧郁气质，他也许深受我们现在所谓的忧郁症之苦。当时的医学认为，黑胆汁——保持平衡才能维持健康的四种“体液”之一——如果占优势就会导致忧郁。事实上，表示黑胆汁的希腊词*melaina chole*正是“忧郁”（melancholy）一词的来源。（同样，被称为多血质、胆汁质和黏液质的性情分别缘于其他三种体液——血液、黄胆汁和黏液——占优势；见第五章。）菲奇诺研究了学术生活与忧郁之间的关联，建议其知识同道改变生活方式以解决问题。菲奇诺制定了一份食谱和药用补品清单，以防体内形成过多的黑胆汁，他的《论从天界获得生命》一文提出用天界的影响来应对这种职业病对学者的危害。

医生认为黑胆汁有冷和干这两种明显性质。土星拥有这些性质，从而与黑胆汁有一种共感关联。因此，任何与黑胆汁和土星相似的东西都要避免。太阳（热—干）和木星（热—湿）的相反性质抵消了黑胆汁的冷—干，因此通过类比扩展，任何与太阳和木星相似的东西都可能有助于缓解学术忧郁。[我们的“快乐”（jovial）一词的字面意思是“与木星有关的”，即显示出这种推理在我们的语言中是多么根深蒂固和得到承认]。因此，为了利用与太阳的共感关联，这位佛罗伦萨人文主义者建议穿黄色和金色的衣服，用向日性的花来装饰房间，获得充足的阳光，佩戴黄金和红宝石，吃“日光”食物和香料（如番红花和肉桂），聆听和歌唱和谐庄严的音乐，焚烧没药和乳香，适度饮酒。然而，当他还建议以古代新柏拉图主义者普罗提诺和扬布里柯为榜样——他将他们的作品从希腊文译成了拉丁文——制作图像来吸引和捕捉行星的力量时，一些读者认为这就有些过分了；对于一个被授以圣职的罗马天主教神父来说，这样做是相当可疑的。事实上，可以把菲奇诺理解成在这一点上跨越了界限，从**自然**魔法步入了**精神**魔法，虽然他很可能对这种解释表示异议。自然魔法运用隐藏于自然中的共感，而精神魔法则求助于精神性的存在——异教希腊哲学中的精灵和诸神，或者基督教神学中的魔鬼和天使。自然魔法不会招致反对，而精神魔法（非常合理地）招致了神学家的谴责。人们针对菲奇诺的正统性提出了一些质疑，但似乎没有采取任何行动，因为可以把这些仪式理解成完全是物理的和药用的，因此完全可以接受。例如，一个多世纪以后，多明我会修士托马索·康帕内拉和教皇乌尔班八世用一场灯火、色彩、气味和声音的仪式（与菲奇诺的处方不无相似之处），来抵消日食期间因暂时失去健康的太阳影响而可能产生的任何不良影响——曾有人预言这次日食会导致教皇死亡。教皇活了下来。虽然在预想的操作中该魔法是自然魔法，但一些旁观者的确认为这样的应用是可疑的。

现如今，对自然魔法的应用以及整个关于共感和类比的内在关联的世界这一观念有时会被斥为非理性或迷信。这种严厉判决是错误的。它源于某种自鸣得意的傲慢和历史认识的匮乏。我们的前人所做的，是对看起来类似的各种神秘自然现象作出观察，并由此将世界中的各种关联和作用传递推广为一种更普遍的说法——一种自然法则。这种推广导出了他们坚持而我们不认同的一个信条，即那些相似或类似的物体正在静静地彼此施加影响。一旦作出这样的假设，体系的剩余部分就可以在此基础上合理地建立起来。他们试图理解世界，试图理解事物并利用自然的力量。他们将观察或叙述的事例归纳成一般原则，然后演绎出其推论和应用。我们也许会说（因为我们知道更新的研究），太阳与向日葵，月亮与大海，或者磁与铁之间的作用，可以不通过隐秘的共感之结而得到更好的解释。但我们并不能说他们的方法或结论是非理性的，或者由此产生的信念和做法是“迷信”。如果允许作这样的跳跃，那么在我们理解世界的过程中最终未被接受的每一种科学理论——无疑包括我们今天相信是对现象的正确解释的一些事物——都将被判定为非理性和迷信，而不单纯是在当时既定的观念、观点和信息条件下通过理性方式得出的**错误**想法。

科学研究的宗教动机

自然魔法只是最强有力地表达了关联的世界、大宇宙和小宇宙以及相似性的力量这些广泛持有的观念。而同样类型的关联和思想往往隐含在从未强调自然魔法的自然哲学家的工作中。例如，当时每一位思想家都确信人、神与自然界之间存在着密切的关联，并因此确信神学真理与科学真理的内在关联。这种特征引出了科学与神学/宗教这一复杂论题。为了理解近代早期的自然哲学，有必要摆脱几种常见的现代假设和偏见。首先，几乎每一个欧洲人，当然也包括本书所提到的每一位科学思想家，都信仰并践行基督教。认为科学研究（无论是否现代）都需要一种无神论——或者美其名曰“怀疑论”——观点，这是那些希望科学本身成为一种宗教的人（他们通常亲自担任圣职）在20世纪提出的一则神话。其次，对于近代早期的人而言，基督教教义并非意见或个人选择，而是自然事实或历史事实。不同教派就更高级的神学观点或仪式活动显然存在着争执，就像今天的科学家就细节进行争论而不去质疑重力的实在性、原子的存在性或者科学事业的有效性一样。神学从未降格为“个人信念”；和今天的科学一样，神学既是一些经过商定的事实，又是对关于存在的真理的不断追寻。其结果是，神学信条被认为是近代早期自然哲学家进行研究所必备的数据集的一部分。因此，神学思想在科学研究和思辨中发挥了重要作用——不是作为外部“影响”，而是自然哲学家正在研究的世界的不可分割的组成部分，需要认真对待。

今天，许多人都会默认那个在19世纪末炮制出来的流传甚广的神话，即“科学家”与“宗教人士”之间进行着一场可歌可泣的斗争。尽管双方的一些成员仍在通过今天的行为不幸地延续着这个神话，但每一位现代科学史家都已经拒斥了这种“冲突”模式，因为它并未反映历史的真实情况。在16、17世纪和中世纪，并没有一个“科学家”阵营在奋力摆脱“宗教人士”的镇压，这些不同阵营根本就不存在。关于压迫和冲突的流行故事充其量是过度简化或夸张，在最坏的情况下则是毫无根据地捏造（见讨论伽利略的第三章）。相反，自然研究者本人都笃信宗教，许多神职人员也是自然研究者。神学研究与科学研究之间的关联部分地基于“两本大书”的想法。圣奥古斯丁和其他早期基督教作家阐述了这种想法，认为上帝以两种不同方式向人类显示自己——一是启示人写出《圣经》，二是创造这个世界即“自然之书”。和《圣经》一样，我们周围的世界是需要进行解读的神圣讯息，敏锐的读者通过研究受造世界可以远为深入地了解造物主。这种深植于正统基督教之中的想法意味着对世界的研究本身就可以是一种宗教行为。例如，罗伯特·玻意耳就认为他的科学研究是一种宗教献身（因此特别适合在周日进行），自然哲学家可以通过沉思上帝的创造而增进对上帝的认识和察觉。玻意耳把自然哲学家描述成“自然的牧师”，其职责就是阐述和解释书写在自然之书中的讯息，收集和表达所有被造物对其创造者的无声赞美。

总之，近代早期的人——以不同方式——看到了一个内在关联的世界，其中一切事物（人类、上帝以及所有知识分支）由千丝万缕的联系构成一个整体。在某些方面，也许可以把生态学和环境科学的新近发展，看成在一定程度上恢复了近代早期自然哲学家在其世界中设想的那个看不见的相互依存网络。无论如何，近代早期的思想家和中世纪先辈一样，凝望着一个充满关联的世界，一个富于目的和意义，饱含神秘、奇迹和许诺的世界。

第三章

月上世界

近代以前，天界依其字面含义差不多占据了人们日常世界的一半。任何人都不可能对天和天的运动视而不见。虽然现代科学对天界运行的解释比过去更好，但现代技术的运用却使大多数人不再能够亲眼看到夜空的运行、感受天界的存在并赞叹它的美，这显得讽刺而有悲剧意味。今人要想像先人那样看到绚丽的夜空，就必须远离光污染和工业污染。早在文字发明很久以前，古人就对天的运动有所认识。然而，解释这些运动却耗费了18世纪之前诸多敏锐天才的心力。对天界隐秘结构的逐步揭示是科学革命的一种关键叙事。那个时代最著名的名字——哥白尼、开普勒、伽利略、牛顿——都是这一叙事中最重要的人物。事实上，在很长一段时间里，天文学的发展代表着科学革命时期的**唯一**叙事，并且是这一时期被称为“革命”的主要原因。

一个生活在公元1500年左右的有识之士会认为，宇宙分为两个区域：**月下世界**包含了从地球到月球以下的一切，**月上世界**则包含了月球及其上面的一切。这一划分出自亚里士多德，他基于日常观察区分了不变的天界和变动不居的地界。在月下世界，土、水、气、火四元素不断地结合、分解和重新结合；新的事物产生，旧的事物消亡。月上世界的情况则完全不同，它是一个不变的区域。在亚里士多德之前数个世纪，观星者看到行星和恒星所走的路径具有完美的规则性。这种变化的缺乏使亚里士多德认为，月上世界由一种同质的东西所构成，即被他称为**以太**的第五种元素，后来的作者称之为第五元素。以太是纯净的基本元素，既不会变化，也不会分解。

观测背景

希腊人开创了一项长远的事业：从物理和数学两方面**解释**天界的运动。这些运动要比今天大多数人所认为的更加复杂和有秩序。每个人都熟知天体的每日升落。天界的一切星体——太阳、月球、行星、恒星——每天升落一次，自东向西穿越天穹。天界的其他运动则要求更加耐心的观测。恒星之所以被称为"恒星"，是因为它们并不相对于彼此运动，而且每隔不到24小时就会回到天空中的同一位置。这就意味着，每颗恒星每晚都比前一晚早升起来一段时间（约四分钟）；因此，你如果在每晚的同一时间观察天空，就会发现，诸星座每晚都会沿着巨大的圆弧缓慢运转；假如你在北半球，这些圆弧的圆心就是那颗永不移动的星——北极星，它位于小熊座的尾端。要想在夜晚的同一时刻看到恒星又回到原先的位置，就得等上一年。由此给人留下的印象是，恒星镶嵌在一个巨大的球壳上，该球壳每隔23小时56分钟绕地球旋转一周。

太阳运行得更慢一些，绕行一周需要整整24小时，这意味着它每天都要改变与恒星的相对位置，**相对于恒星背景自西向东**缓慢运行，需要一年时间才能回到同一颗星的附近。月球的运动与此类似，但要明显得多。它每晚比前一晚**迟**升起50分钟，因此你如果接连几晚在同一时间寻找它，就会发现它每晚都向东走了一段距离（不妨试试！）。29天后，月球又回到了初始位置。

行星的运行也大同小异，但路径更为曲折怪异，这强烈吸引着人们去寻求解释。在大多数时间里，行星就像太阳和月球一样，相对于恒星背景自西向东缓慢移动。但每隔一段时间，行星就会慢下来，停住不动，转而朝相反方向自东向西运行。这种现象被称为**逆行**。再过一段时间，行星再度停下来，掉转方向继续常规的运动。

古希腊人把太阳、月球、水星、金星、火星、木星和土星这七个看起来相对于固定的恒星背景移动的天体称为“行星”（意思是“漫游者”）。但行星不会漫游得太远，它们的运动局限在天上狭窄的黄道带中。黄道被分为等长的十二段，每一段都包含一个星座或“宫”，比如白羊座、金牛座、双子座等等。于是，随着诸行星相对于恒星背景作各自的运动，它们就好像沿着黄道带从一个宫运行到下一个宫。一个人所属的“宫”就是他出生那天太阳“所在”的黄道宫。我们很快会讨论更多有关占星学的内容。

历史背景

柏拉图确信，天界是按照和谐的数学法则运行的。他的灵感来自于毕达哥拉斯学派的观点，该学派是一个秘密宗教团体，认为数学——数、几何图形、比例与和谐——同时是宇宙和有序生活的真正基础。对于柏拉图和近代以前受他影响的人而言，造物主是一位几何学家。然而，行星的不规则运动似乎与一个有序数学世界的观念相悖。因此柏拉图声称，行星的运动仅仅**看上去**是不规则的，我们凭借肉眼无法看到其背后的神圣规律。由于柏拉图认为圆是最为完美和规则的形状，圆周运动是无始无终的从而是永恒的，因此他要他的学生用**匀速圆周运动**的组合来解释行星的可见运动。这一要求启发着两千多年来的天文学家。

柏拉图的学生欧多克斯提出了一种宇宙模型，它由以地球为中心的一系列同心球（宛如一层层洋葱皮）所组成。每个天球匀速旋转，但每颗行星都会获得若干天球的运动，这些运动组合起来（大致）就是行星的视运动。欧多克斯体系是一个**数学**模型。他并不关心天界在物理上如何运作，也不在乎天球是否真的存在，关键是用数学来说明现象。而亚里士多德则试图建立一个**物理**模型。他把欧多克斯的天球变成了实在的坚固物体，这些天球实际携带着行星旋转；他还解释了运动如何像天界机械装置的齿轮一样从一个天球传到下一个天球。亚里士多德的功绩在于将天文学和物理学和谐地结合起来（图2）。

同心球模型的问题在于不能精确地解释天文观测，例如行星的亮度会发生变化，就好像它们时近时远，四季长度也不尽相同。这一切都无法用以地球为中心的同心球模型来解释（图3）。

后来的天文学家试图解决这些问题，其顶峰是托勒密（约90—约168）的体系。为了解决季节不等的问题，托勒密引入了**偏**

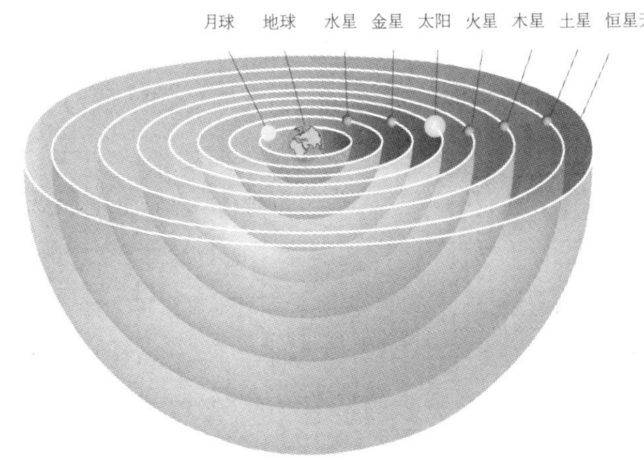

图2　亚里士多德同心球模型简化版本的剖面图

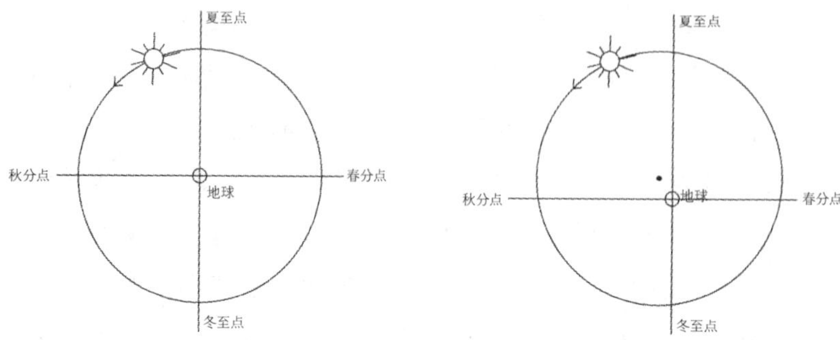

图3　（左图）假如地球位于太阳天球的中心，太阳的周年视运动将被分成四段等长的弧，使四季等长。但事实上夏天比冬天更长；（右图）托勒密的偏心地球模型将太阳的轨道分成四段不等的弧，对应于长度不等的四季。这种安排还解释了为什么太阳在夏天似乎移动得更慢：因为那时太阳离地球更远。

心圆：也就是说，他把地球移出了中心。在他的体系中，每一个天球都有自己的中心，其中没有一个与地球重合。

为了更好地说明行星的位置并解决行星亮度变化的问题，托勒密引入了**本轮**（图4）。每颗行星都沿一个小的圆形轨道运行，轨道中心在一个环绕地球的大圆（均轮）上运动。本轮和均轮的运动组合极好地解释了行星表观的环圈路径，行星在运动过程中有时会靠近地球，因而显得更亮。

托勒密体系能够很好地预言行星的位置，但它更多地是一个数学模型而非物理模型。亚里士多德物理学认为重物会落向宇宙的中心，因此球形的地球位于宇宙中心，重物会下落。但托勒密模型中的地球不在中心，它为什么不会移向中心呢？重物为什么会落向宇宙中心之外的某个地方？数学模型与物理体系之间的这种不一致困扰着中世纪的阿拉伯学者，而在当时的欧洲，

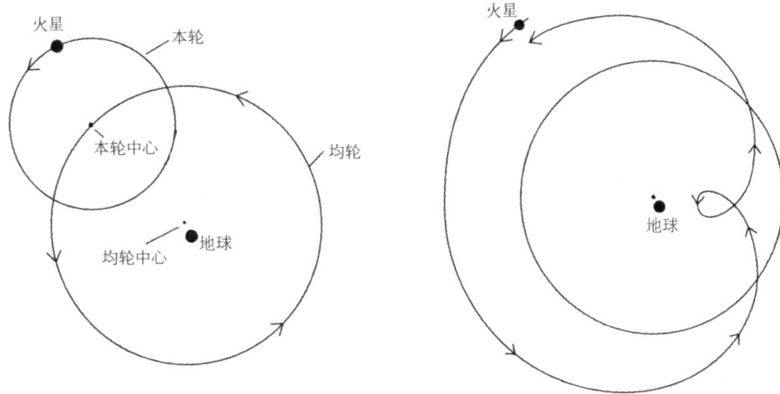

图4 （左图）托勒密为行星设计的本轮和均轮。行星在本轮上（从地球的北极俯视）逆时针运行，同时本轮在均轮上也作逆时针运动；（右图）由本轮和均轮运动合成的行星视运动。行星位于均轮外侧时显得暗一些，并且自西向东运行；行星位于内侧时因为更近而显得亮一些，最靠近地球时会自东向西运行（逆行）。

亚里士多德和托勒密的工作还不为人知。伊本·海塞姆（或称阿尔哈增，约965—1040）采取了一个折衷方案。他的体系含有以地球为中心的天球，这会使物理学家感到满意。但这些天球坚实而有厚度，足以容纳不以地球为中心的环状通道，行星在这些环状通道中沿着本轮和均轮运行，从而解释了观测到的现象（图5）。

中世纪的欧洲天文学家继承了这些观念和问题，和他们的阿拉伯同行一样继续完善和更新这一体系，以求最精确地预言行星的位置，偶尔也会试图构建一个在物理上令人满意的体系。

THEORICAE NOVAE PLANETARVM GEORGII PVRBACHII ASTRONOMI CELEBRATISS.

DE SOLE

Ol habet tres orbes a se iuicē omniquaqꝫ diuisos atqꝫ sibi cōtiguos Quoꝶ supræ/mus secūdū superficiē conuexā est mūdo cōcentricus:secūdū cōcauā aūt eccētricus Infimns uero secūdū cōcauā cōcentric⁹: sed secūdū conuexā eccētric⁹ Tertius aūt ī hoꝶ medio locatus tam secūdū super/ficiem suā conuexā qꝫ concauā est mūdo eccentric⁹.Dicit̃ aūt mūdo cōcētric⁹or/

THEORICA ORBIVM SOLIS.

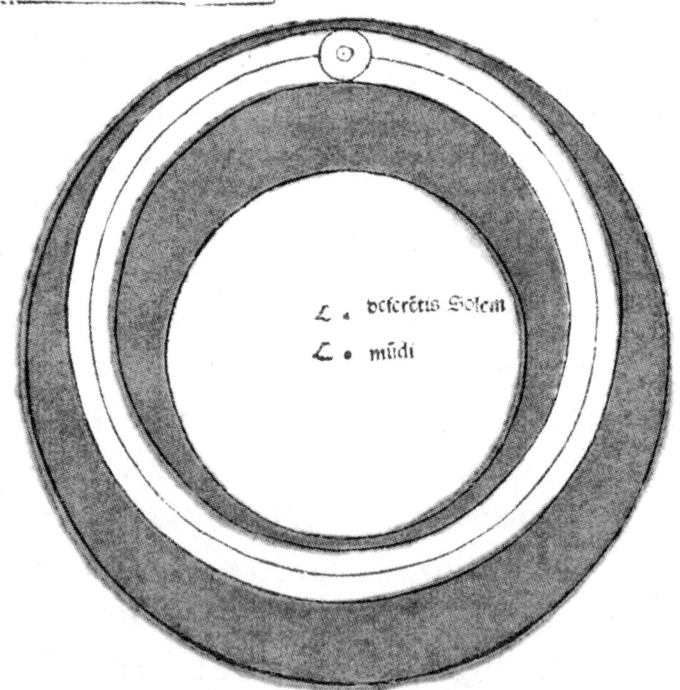

图5　普尔巴赫所普及的对伊本·海塞姆有厚度的天球模型的改进，它进入了15世纪的天文学标准教科书——萨克罗博斯科的《天球论》及后来的版本。该图取自1488年的威尼斯版，描绘的是太阳天球。

近代早期的天文学模型

尼古拉·哥白尼（1473—1543）一生中的大部分时间都在担任弗劳恩堡（今波兰境内的弗龙堡）大教堂的教士，这是一个行政性质的圣职。他曾在博洛尼亚大学学习教会法，在帕多瓦大学学习医学，1503年在费拉拉大学获得法学博士学位。在博洛尼亚期间，哥白尼开始研究天文学，到了1514年左右，他写了一份思想概要，声称行星系统的中心不是地球，而是太阳。在他的**日心**体系中，地球每日绕轴自转一周，这产生了人们所熟悉的一种表象，即整个宇宙绕地球旋转。太阳沿黄道的运动实则是一种假象，其真正原因是地球的绕日运动。观察所见的火星、木星、土星的“环圈路径”和逆行并非缘于它们自身的运动，而是缘于**我们**地球的运动**与它们**各自绕日运动的叠加（图6）。只有月球是绕地球运转的。

哥白尼的工作以手稿形式流传，这足以确立他作为天文学家的声誉。1515年，教会的一个委员会希望改革从罗马时代沿用下来、需要彻底改变的旧儒略历，于是写信征求哥白尼的意见。（哥

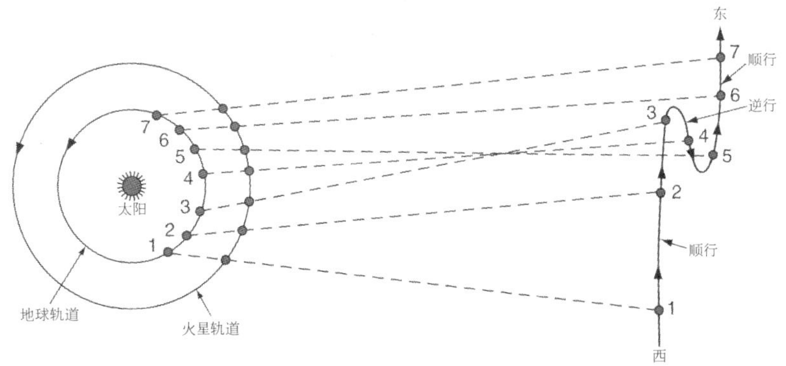

图6　哥白尼对某颗“更高的”行星即外行星（火星、木星、土星）逆行运动的解释。当地球行经其中一颗外行星时，就会造成“环圈路径”的假象。

白尼回复说，首先需要更加精确地确定太阳年的长度。）然而，哥白尼并未发表自己对天文学体系的完整阐述。在超过25年的时间里，他一直在完善该体系，要不是几位显要的教士催促他，其成果可能永远都不会发表。例如，1533年，教皇的私人秘书维德曼施泰特讲述了哥白尼体系，教皇克雷芒七世和一些红衣主教听后甚为高兴。卡普亚的红衣主教舍恩贝格写信给哥白尼：

> 我听说您主张地球在运动；太阳的位置最低，因而是宇宙的中心，……还听说您为这一整套天文学体系给出了说明，……因此我强烈恳请您让学界知晓您的发现。

然而哥白尼依旧含糊其辞，忙于大教堂教士的职守，表示害怕别人批评他的体系过于新颖。

1538年，维滕贝格大学的梅兰希顿派年轻的天文学教授格奥格·约阿希姆·雷蒂库斯来哥白尼这里学习。雷蒂库斯编写发表了一份哥白尼思想的概要，反响很好，哥白尼终于同意发表其完整的手稿，并交由雷蒂库斯出版。雷蒂库斯接手了这项任务，但不久以后他在莱比锡找到了一份工作，遂把出版之事交由路德教牧师奥西安德尔负责。奥西安德尔完成了出版工作，《天球运行论》终于在1543年问世——哥白尼在临终前看到了这本书。

该书的问世并没有招来哥白尼担心的那种批评。不少人读了它，但几乎没有人真正相信。直到16世纪末，坚定的哥白尼主义者可能不过十几人。这是为什么？因为哥白尼的日心体系并不比地心体系更好地符合观测数据，在物理上也没有更简单。事实上，为了让他的体系与观测相符，哥白尼不得不保留本轮，并且让太阳偏离中心。更严重的是，地球运动的观念与基本的物理学、常识乃至可能与《圣经》相抵触。像地球这样的天体自然会落到最低处，即宇宙的中心——这一“自然位置”原理解释了重物为什么会下落。那么，整个地球如何可能悬在离中心这么远的地方呢？常识表明我们并没有在运动。为了每天转一圈，地球必须转得很快，但我们对这种运动浑然不觉，飞鸟和云朵也没有因为地球在其下方高速旋转而落在后面。一些中世纪思想家曾经探讨过地球旋转的可能。尼古拉·奥雷姆（约1325—1382）断言，所有运动都是相对的，如果没有参照点就不可能确定究竟是地球在旋转还是天在旋转。他最后总结称，地球静止而天界运动似乎更有可能。不少对《圣经》作字面解读的人会援引那些说地球静止而太阳运动的段落，尽管解释各不相同。最后，假如地球绕太阳运转，恒星应当有视差——随着地球从轨道的一侧转到另一侧，恒星的相对视位置应当有微小的改变。但在当时，人们从未观测到视差，这就表明，要么地球**不在**运动，要么那些恒星远得**无法想象**。13世纪时，诺瓦拉的康帕努斯估计土星天球（恒星就在这层天球上方）的高度约为7300万英里，即使是游历最广的中世纪人也会对这一距离感到震惊。哥白尼估计土星天球的高度约为4000万英里，但是根据后来的计算，恒星和我们的距离至少要有1500**亿**英里，才会观测不到恒星视差。如此广袤的虚空在哥白尼时代的读者看来是荒谬绝伦的。（事实上，连最近恒星的距离也是由观测不到恒星视差所最保守估算的距离的170倍。恒星视差直到1838年才被发现。）

即使没有观测上的证据，也有几个因素能使哥白尼确信日心说。在致教皇保罗三世的献词中，哥白尼把托勒密体系，包括其偏心圆、本轮和对每颗行星的单独处理，说成是一个“怪物”。既然世界“由最高超且最有条理的工匠所造”，它理应是和谐的。作为人文主义者的哥白尼认为自己是在清理后来的“添加”，以回到柏拉图关于揭示井然有序的天界运动的原初召唤。由于担心自己的体系过于“新颖”，哥白尼通过援引古代的先驱——阿里斯塔克、毕达哥拉斯以及西塞罗所提到的希克塔斯，来尽可能地减少显示出来的新颖性；哥白尼甚至对《圣经》中的某些段落作了重新阐释以支持日心说。

然而，人们完全可以在欣赏哥白尼体系的同时又并不相信其真实性。在日心体系中，确定行星位置的星表更容易计算；因此，一些天文学家把它当作一种“方便的虚构”而加以接受。哥白尼本人将日心说视为对世界的真实描述，但奥西安德尔在哥白尼的书中偷偷加入了一篇（未署名的）序言，从而削弱了它的效力。奥西安德尔写道，我们“对行星运动的真正原因一无所知”，而且：

> 这些假说无须为真，甚至也并不一定是可能的；只要它们能够提供一套与观测相符的计算方法，那就足够了……谁也不要指望能从天文学中得到任何确定的东西，因为天文学提供不出这样的东西。他也不该把为了其他目的而提出的想法当作真理，以便在离开这项研究时比刚刚开始进行研究时更为愚蠢。

即使哥白尼当时没有中风，他读到奥西安德尔的序言时多半也会。雷蒂库斯大为光火，把自己书上奥西安德尔的序言撕掉了。数学模型与物理体系之间的张力又一次展现出来。多数天文学家主要对行星在某一时刻处于哪个位置感兴趣；至于究竟是太阳绕地球转，还是地球绕太阳转，这根本无关紧要，许多人怀疑是否真的有人能够确定何者为正确。对于天文学理论而言，只要能正确地计算出行星的位置、给出星表，这就够了。对于大多数人而言，实际结果比理论更重要。要想理解这一点，我们需要意识到，早在托勒密时代之前，天文学研究背后的主要驱动力就一直是占星学，而占星学是很实际的，需要计算出行星在多年以前或多年以后精确到分的位置。

实践天文学，或占星学

天文学（字面意思是"星体的法则"）测量和计算天体的位置，并提出假说性的宇宙论体系；而占星学（字面意思是"对星体的研究"，与地质学、生物学等类似）则致力于解释和预言天体对地界的影响。一般来说，这两项事业——前者是理论的，后者是实践的——是由同一批人从事的。近代早期的许多天文学家都主要以从事占星学为生。不要把古代、中世纪或近代早期的占星学与"报纸上根据天宫图算命"的无稽之谈混淆起来。占星学是一项严肃而精深的活动，其基本想法是天体会对地界产生某些影响——这是关联世界观的关键一环。中世纪和近代早期的大多数占星学并不是"魔法的"、超自然的或非理性的，而是依赖于作为世界组织方式之一部分的自然机制。既然光能从行星传到我们这里，为什么不能有一些别的影响伴着光传来，就像火光也能加热远处的物体一样？天界对地界的影响很容易观察到——月球与潮汐相关联，太阳在黄道上的位置决定了季节气候。对人体的影响也同样明显，例如月球周期与人的月经同步。天界影响的真实性十分明显，以至于毋庸置疑；有关占星学的诸多争论其实涉及的是这些影响的程度，以及如何准确预测其效果。七颗行星时刻改变着彼此的相对位置（"星位"），在黄道十二星座（这些星座自身也在不停地穿越十二个"宫"，即相对于地平线的位置）之间往来穿梭，来自这七颗行星的交叉影响形成了一个极为复杂的系统。这一系列征象与禁忌、已知与未知，其复杂性完全不亚于如今人们对全球气候变化因素的探究或是对未来经济走向的预测。与后者相比，近代早期占星学家的成功率或许还会高一些。

占星学包括几个相互交叠的分支。气象占星学致力于预测来年的天气。许多从业者往往被径直称为"数学家"，这表明占星学需要计算；这些人以编写历书为生，书中包含了历法、月球周期、日月食的时间、对天气的预测（就像今天的《农民历书》一样）以及对重要事件或趋势的预言。印刷术使占星学作品变得廉价易得，传播广泛。医生们借助医学占星学来确定治疗过程的关键时间，并提出疾病的可能病因（见第五章）。本命占星学依据一个人的出生地点和出生时的行星位置来确定行星"印入"新生儿的影响。行星影响的特定组合会在体液系统中产生独特或天生的"体质"，这导致了特殊的倾向和特征。这些倾向（易于罹患某些疾病、发怒、懒惰或忧郁等等）可能因为后来的行星排列而暂时加强。因此，这种占星学旨在获知一个人天生的体质，以了解其特殊的长处和弱点，提醒注意可能发生危险或有益健康的时间。这种活动的更强形式渐渐变为一种神判占星学，其决定论色彩（即认为星体的影响支配着我们的行为和命运）令人无法接受，因而广受批评。神学家们谴责这种观念违背了人的自由意志。近代早期学术界的共识是，"星体影响但并不强迫"我们，"有智慧的人支配星体"（*sapiens dominatur astris*）。简而言之，人总能选择行动，尽管完全自由地行使意志可能受制于外部影响（例如，火星的特定位置导致体液失衡，进而使人一时冲动，理性能力减弱）。事实上，近代早期的占星学与现代的"先天本性与后天培育"之争有类似之处，它们都试图去解释人的行为。具有讽刺意味的是，其显著差异在于，现代人似乎忘记了自由意志的首要性。

神判占星学有时被用来确定重大事件的良辰吉日。数学家兼魔法师约翰·迪伊（1527—1608/1609）便用占星学为伊丽莎白一世选择加冕吉日。最早的科学社团之一猞猁学院的成立日期是根据天宫图选定的，新的罗马圣彼得教堂的奠基日也是如此。有时，根据占星学选择日期并不是要获得什么有利的“影响”，而是要为事件增添意义，一如美国科学家特意让火星探测器在美国独立日那天着陆。神判占星学还被用来预言未来的事件，比如战争和死亡，这潜在地远离了**自然**因果性，而后者正是近代早期学术占星学的基础。要想解决这个问题，可以把一些天象（尤其是彗星）看成**征兆**而非**原因**，当作神所传达的有关未来之事的迹象。对天界征兆的兴趣在新教盛行的北欧更加明显，这部分是由于梅兰希顿为萨克罗博斯科《天球论》（一本天文学基础教材）的新教版本所作的一篇序言。他在序言中强调了占星学对于理解上帝在天界中的迹象的重要性。总之，各种类型的占星学为更好的生活提供了有用信息；它在近代早期思想中无处不在，这表明月上世界的确是人们日常世界的一半。

天界变化与神圣和谐

伴随着对天界征兆的占星学兴趣，丹麦的贵族天文学家第谷·布拉赫（1546—1601）首次登场。1572年11月，他发现了仙后座有一个明亮物体，而那里本该什么也没有。第谷惊讶不已——那个物体是什么，意味着什么？第谷在1573年的占星历书中试图作出解释，断言它预示着即将到来的骚乱和动荡。第谷观察这个明亮的光点，它并不像彗星那样会移动。第谷和其他欧洲天文学家试图测量它的周日视差，希望由此推算出它的距离，但他们并没有观测到视差，这意味着此物体比月球远得多，亦即处于月上世界——人们一向认为这个世界是没有变化的，但它却是一颗**新**星。[第谷看到的是一颗超新星；那次猛烈爆发的不断膨胀的残留部分在1952年被探明。“新”（nova）来自第谷用来指该星体的拉丁词——新星（*stella nova*）。]

不久以后的1577年，天空中出现了一颗明亮的彗星。亚里士多德曾经教导说，彗星和流星一样是月下世界的现象，源自上层大气中散发的可燃物的燃烧。彗星是游移变化之物，不可能处于不变的月上世界。第谷在占星学上断言，1577年的彗星显示了与那颗新星相同的预兆，但这一次他观测到了彗星的周日视差。第谷的观测结果得到了他人确证，该结果表明这颗彗星位于远在月球之上的金星天球。1585年，当另一颗明亮的彗星出现时，第谷给出了相同的观测结果。这些彗星进一步表明，“不变的”天界存在着变化，彗星的位置变化则表明它们正在**穿越**行星天球，这意味着带着行星运动的坚实天球并不存在。那么，是什么使行星沿着规则的路径运行呢？行星如何能够摆脱坚实的天球，这着实令人困惑，但它意味着天体的路径能够彼此交叉，这使第谷设计出一个新的天界体系，将其观测结果与托勒密和哥白尼体系中的精华部分结合起来，同时又避免了两者之中招致非议的部分。在第谷的**地日心**体系中，地球如常识和《圣经》所指示的那样静止于宇宙的中心，月球则围绕地球运转。而行星则围绕太阳运转，太阳带着诸行星围绕地球运转。

第谷在丹麦海峡的汶岛上建造了城堡式的天文台——天堡，在那里继续观测天空，其准确程度前无古人。此时，约翰内斯·开普勒（1571—1630）这位坚定的哥白尼主义者正在写下自己的惊人发现。16世纪90年代，开普勒在格拉茨的一所高中教书时苦苦思索着现代科学家根本不会去问的一个问题。在哥白尼体系中，围绕太阳运转的行星只有六颗，而不再是围绕地球运转的七颗。而假如有七颗行星，它们就能与一周的七天、七种已知金属、音阶的七个音以及世界上所有其他重要的“七”很好地相合。七颗行星有一种美妙的和谐，适合于一个相互关联的世界；而六颗行星就不行。那么为什么有且仅有六颗行星？上帝又为什么恰好将它们置于如今那么远的距离呢？近代早期的世界中充满了意义与目的，其中一切事物都有某种讯息需要解读。

开普勒在1595年7月19日的课堂上突然意识到，假如在圆内作一个内接正多边形（如正三角形、正方形、正五边形等等），再在这个正多边形内作一个内切圆，那么我们就得到了两个圆，其相对大小是由正多边形的种类决定的。兴奋不已的开普勒开始计算由不同正多边形所决定的比值，看其中是否有某个比值与行星到太阳的距离之比相符合。但他失败了。不屈不挠的开普勒又用球体和正多面体代替了圆和正多边形。这一次，通过将球体和正多面体以适当顺序套起来，开普勒得出，球体的相对大小符合哥白尼理论所给出的行星到太阳的相对距离。不仅如此，由于正多面体（所有面全等的五种所谓的柏拉图立体，即正四面体、立方体、正八面体、正十二面体和正二十面体）只有五种，因此被它们隔开的球体**有且只有六个**，从而行星的数目不多不少就是**六颗**。对于开普勒而言，这一发现令人敬畏。他找到了行星数目和距离何以如此的原因，揭示了天界的几何结构，其优雅和美正是哥白尼体系的最佳证明。这一惊人关联不可能出于偶然；开普勒发现了上帝创造天界时所使用的数学方案。

开普勒例证了近代早期典型的人类探索的统一性。神学研究与科学研究并不截然分离：研究物理世界意味着研究上帝的创造物，研究上帝则意味着了解世界。其实，开普勒之所以确信哥白尼的学说，部分是因为日心宇宙为三位一体提供了物理上的对应：位于中心的太阳象征圣父，接收和反射太阳光的恒星天球象征圣子，两者之间充满光的空间则象征圣灵（在神学上代表圣父和圣子之间的爱）。开普勒及其同时代人援引自然之书和《圣经》这两本大书的观念，确信上帝在创造的世界中植入了有待人们发现的讯息。于是，在自然之书中解读出讯息这一神学动机为整个近代早期的科学研究提供了最大的驱动力。

开普勒在《宇宙的神秘》（1596）一书中宣布了他的发现，并且给第谷·布拉赫寄了一本。第谷邀请开普勒与之合作，开普勒起初谢绝了，但在第谷到鲁道夫二世皇帝在布拉格的宫廷担任皇家顾问之后，开普勒于1600年投奔了他。第谷于次年去世，鲁道夫二世让开普勒接任皇家数学家一职。第谷曾让开普勒研究火星的运动，开普勒长期竭力寻找一条与第谷的观测结果相符的轨道，最终得出了一个惊人的结论。他发现，只有让火星沿着**椭圆**轨道而不是圆轨道运动，火星的位置才能得到最好的解释。这样一来，开普勒不得不无奈地与两千年来以圆周为基础的天文学传统决裂。不过，既然（用开普勒的话说）第谷"打碎了水晶天球"，又是什么把行星维持在椭圆轨道上运行呢？开普勒假定太阳之中有一种"致动灵魂"（*anima motrix*），这是一种能够推动行星的力量。和太阳光一样，这种力量随着距离的增大而衰减，因而距离太阳越远，行星运动就越慢。开普勒援引威廉·吉尔伯特（1544—1603）有关地球是个巨大磁体的新近断言（见第四章），假定太阳能够发出第二种力量，它与磁力类似，在某些地方吸引行星，在另一些地方排斥行星。致动灵魂与"磁"性共同作用，使行星无须天球带动就能沿着椭圆轨道运行，被拉近太阳时运动得快些，被推远时运动得慢些。虽然开普勒放弃了匀速圆周运动，但他兴奋地发现了另一种均匀性——“等面积定律”——来取而代之，即当行星运动时，太阳与行星的连线在相等时间内扫过相等的面积。同样，虽然开普勒协助瓦解了亚里士多德的宇宙，但他仍在《哥白尼天文学概要》一书的副标题中指明它是亚里士多德《论天》的“补遗”。连续与变化兼有、创新与传统并存正是近代早期自然哲学的典型特征。

望远镜和地球的运动

第谷用裸眼观测天象的能力无人能及，而他也是最后一批这样做的人之一。当开普勒埋头计算时，伽利略·伽利莱（1564—1642）听说荷兰人发明了一种能使远处物体显得更近的仪器，便亲自作了改进，并于1609年将其指向了天空。几乎每把镜筒（*occhiale*，后称作望远镜）指向一处，伽利略都会有新的发现。他发现月球表面布满了山脉、峡谷和海洋——换言之，月球看上去与地球非常相似，因此也是由四元素而非亚里士多德所说的第五元素组成的。他发现木星周围有四颗卫星，宛如一个小的行星系统。伽利略根据托斯卡纳大公科西莫二世·德·美第奇之名将这些卫星命名为“美第奇星”，从而使自己名利双收。伽利略发现土星的形状很奇怪，看起来就像是三个连在一起的球体。他还发现金星像月球一样会显示出位相。最后这项发现第一次有力地反驳了托勒密体系，因为在托勒密体系中，金星总是位于太阳与地球之间，因而最多只能呈现出新月形。伽利略观测到了新月形的金星**和**满的金星，这表明金星必定时而处于我们和太阳之间，时而远在太阳的另一侧，简而言之，金星围绕太阳运转。从此以后，天文学家只能在第谷体系与哥白尼体系之间进行选择（图7）。于是，这两个体系之间唯一的分歧，即地球是否运动，就成了天文学家最为关切的问题。

图7　里乔利《新天文学大成》(1651)卷首象征性的插图比较了三种世界体系。正义之神阿斯特莱亚正在权衡哥白尼体系和里乔利体系(对第谷体系作了些微调整),托勒密则斜倚在自己已遭抛弃的体系那里。画面上方的小天使手拿行星,显示出新近的发现:水星和金星的位相、月球的粗糙表面、木星的卫星以及土星的“把手”。上帝之手赐福于世界,伸出的三根手指旁标着“数、重量、量度”(《智慧书》11 : 20),象征着造物的数学秩序。

伽利略迅速出版了《星际讯息》一书，公布了他用望远镜获得的第一批发现，并将这本书与望远镜一道寄给了全欧洲的天文学家和统治者。许多人难以看到伽利略所描述的现象，因为望远镜的放大倍数不高，光学系统很差，难以使用。罗马的耶稣会天文学家提供了关键支持，他们证实了伽利略的观测结果并且继续作出观测，还在1611年举办盛宴向伽利略表示敬意。罗马学院的资深成员克里斯托弗·克拉维乌斯（1538—1612）是欧洲最受尊敬的数学家之一，他设计了教皇格里高利十三世于1582年开始施行（持续至今）的格里高利历；他写道，伽利略的发现要求人们重新思考天界的结构。虽然克拉维乌斯和许多其他人坚持地心说，但一些年轻的耶稣会天文学家可能转向了日心说。然而，这些友好的交往没有继续下去，因为伽利略与两位耶稣会天文学家发生了争论（他在其中常常显得无礼）：一是与克里斯托弗·沙伊纳争夺发现太阳黑子的优先权并且争论黑子的本性，二是与奥拉齐奥·格拉西争论彗星（格拉西支持第谷的观测，认为彗星是天体，而伽利略坚持说彗星是月下世界的视觉幻觉）。

在科学史上，“伽利略与教会”这一情节变成了最大的神话，遭到了最严重的误解。这些事件缘于一系列纠缠不清的思想、政治和私人因素，它们极为错综复杂，历史学家至今仍在试图厘清。这并不单单是“科学与宗教的冲突”问题。教会内外兼有伽利略的支持者和反对者。与事件有密切关系的因素包括：情感受到冒犯、政治上的阴谋、解释《圣经》的资格、未占天时地利以及被不同教派裹挟。最后一个引爆因素是，伽利略于1632年出版了《关于两大世界体系的对话》，这本书对托勒密体系和哥白尼体系作了比较，并且明确把后者当成正确的，声称地球在运动。伽利略的主要证据是，他认为地球的运动引起了潮汐；在这一点上他完全错了，尽管说地球在运动是正确的。究竟哪一个体系是正确的，与教会本无直接利害关系；地心说和亚里士多德主义都不是教会的教条。然而，《圣经》解释的确与教会休戚相关，不仅地球运动对解释《圣经》有所暗示，而且伽利略在17世纪10年代初为了支持自己的观点更是莽撞地涉足其中。这种对《圣经》的随意解读就像是当时的新教为了拒斥传统解释而许可教徒作出对自己有利的解读。结果，1616年教会要求伽利略把日心说和地球运动的观点当作假说，在找到强有力的证据之前不得认为它们真正正确；伽利略同意了。1624年，伽利略从他的朋友、当时已是教皇乌尔班八世的马费奥·巴贝里尼那里得到了写作《对话》的许可，条件是伽利略要在书中申明教皇在方法论上的观点，即自然现象（如潮汐）可能有若干种原因，其中一些是不可知的，因而我们不能绝对确定地把现象归于某单一原因。伽利略照做了，但只是借一个自始至终扮演傻瓜的角色之口在书的末尾表达了这种观点。伽利略还“忘了”把自己1616年答应教会的事告诉乌尔班。该书（经教廷审查人员的许可）出版后，一切大白天下，乌尔班大发雷霆，感觉自己受了欺骗和羞辱。更糟的是，乌尔班正被当时关于三十年战争的外交谈判、日益强烈的批评、废黜他的阴谋以及有关他即将死去的谣言弄得焦头烂额，原本并不足道的恼火被激化了。宗教裁判所为伽利略拟了一份认罪辩诉协议，建议把伽利略遣送回家并稍作惩罚，但盛怒之下的教皇拒绝了，他坚持要严惩伽利略以警戒他人。伽利略被要求发誓放弃地动说（伽利略照做了），其著作也被教会查禁。值得注意的是，几位红衣主教，包括乌尔班的侄子，都拒绝在对伽利略的判决书上签字。伽利略从未如民间传说中那样被判为异端，遭到关押或囚禁。

最终，伽利略被判软禁于他在托斯卡纳山的别墅中。他在那里继续工作、教学，并且写出了或许是他最重要的一本著作——《两门新科学》。很难说教会对伽利略的判决究竟产生了多大影响。一方面，它使一些自然哲学家对自己的哥白尼主义信念三缄其口。例如，听到伽利略受谴责的消息后，勒内·笛卡尔（1596—1650）将一本刚刚完成的支持日心说的著作藏了起来。像耶稣会士这样担任天主教圣职的人如今不再能够公开支持哥白尼的学说，因此转向了第谷体系或其变种（图7），尽管有时是阳奉阴违。另一方面，在意大利和其他天主教国家，包括天文学在内的科学研究仍然继续着，尽管有时需要回避敏感话题。

继前两代人的观念剧变之后，17世纪中叶的天文学进展更多是在观测和技术方面，而非理论方面。法国神父皮埃尔·伽桑狄（1592—1655）于1631年第一次观察到了水星凌日，1630年辞世的开普勒曾经预言过该现象。改良后的望远镜带来了新的发现和更准确的测量，但由于需要避免球面像差和色差，望远镜造得越来越长、越来越笨重，有的甚至长达60英尺。不过这样人们就能发现土星的奇特形状是一系列环，1656年，克里斯蒂安·惠更斯（1629—1695）发现了土星最大的卫星。吉安·多梅尼科·卡西尼（1625—1712）在巴黎借助罗马光学仪器商朱塞佩·康帕尼制造的精良望远镜又发现了土星的四颗卫星，并依照路易十四的名字将其命名为卢多维奇星。耶稣会士乔万尼·巴蒂斯塔·里乔利（1598—1671）编制了新的星表，并和他的同行弗朗西斯科·玛里亚·格里马尔迪（1618—1663）合作绘制了一张详细的月面图，图上许多有特色的名字一直沿用至今，其中包括以哥白尼来命名最为突出的几座环形山之一。在格但斯克，约翰·赫维留（1611—1687，或许是最后一个同时用肉眼和望远镜进行认真观测的人）也画了一张月面图，他还观测了彗星，参与了全欧洲有关彗星是沿直线运动还是绕日运转的讨论。

行星不借助于坚实的天球如何能够沿恒定轨道运动，这个问题继续引人思索。笛卡尔提出了一个无所不包的世界体系，成为17世纪最重要的体系之一。笛卡尔设想整个空间都被不可见的物质微粒填满。这些微粒一刻不停地在环流或涡旋中运动着。我们的太阳系就是一个由这些微粒构成的巨大涡旋，像水中漩涡带着稻草一样裹挟着行星旋转。这种涡旋模型简洁地解释了为什么行星都沿同一方向且几乎在同一平面上运行。地球本身处于一个较小涡旋的中心，该涡旋带着月球在轨道上运行，地球周围的物质漩涡形成一股"风"将物体推向地心，这就解释了重力现象。笛卡尔的涡旋理论为天体的运动提供了一种综合解释，在通俗讨论和教科书中流传甚广，但它太不精确，对天文学家没有实际价值。

牛顿（1643—1727）年轻时曾拥护笛卡尔的涡旋理论。17世纪60年代初在剑桥读书时，牛顿研究了当时多数大学里仍是本科生标准教科书的亚里士多德著作。但牛顿很快就开始在课外阅读像笛卡尔这样的"现代人"的思想。牛顿接受了笛卡尔解释行星运动和重力的原理的一个修改版本。但是到了17世纪80年代初，牛顿的想法开始转变。他抛弃了笛卡尔的涡旋，开始构想太阳与行星之间存在着一种吸引力。这种想法有几个来源，特别是人们所熟知的磁现象以及开普勒曾经假定的太阳与行星之间存在的"类磁"力。对于开普勒而言，正是这种"磁力"与致动灵魂的结合使行星沿椭圆轨道运行。而在牛顿这里则是惯性（行星沿轨道切线运动的倾向）与朝向太阳的吸引力（我们称之为引力）之间的平衡产生了稳定的椭圆轨道。伦敦皇家学会的几位会员曾以类似的思路解释过行星的运动，特别是罗伯特·胡克（1635—1703）曾在1679至1680年写信给牛顿谈了自己的想法。后来胡克抱怨牛顿剽窃了自己的想法而没有给他以足够的尊重，这使得神经极度敏感的牛顿在自己的著作中对胡克只字不提，并且终生将胡克视为宿敌。牛顿发表于《自然哲学的数学原理》（1687）中的伟大成就，用纯数学的方法重新导出了开普勒根据第谷的观测数据经验得出的行星运动定律，并使引力变得真正普遍，即任何两块物质之间都存在引力。开普勒无疑会对此感到欣慰，因为现在有更多证据表明上帝是按照和谐的数学方案创造了世界。牛顿的万有引力定律最终消除了以往关于地界物理学与天界物理学之区分的最后一抹痕迹——行星运转与苹果下落服从的是同一个定律。

并非所有人都为此欢欣。通过复兴引力的观念，牛顿似乎使一种沉寂了约70年的想法死灰复燃。一种不可见、非物质的力可以在一切物体之间发生作用而没有任何机制或明显原因，这不仅比物质性的笛卡尔涡旋更难以理解，而且在许多人看来是又回到了亚里士多德主义者所说的“隐秘性质”或自然魔法中的共感。事实上，17世纪下半叶自然哲学的前沿问题始终是用不可见微粒的运动来解释那些看起来是吸引或共感的现象（见第五章）；如今牛顿似乎是在开倒车。

牛顿曾与之争夺微积分发明优先权的戈特弗里德·威廉·莱布尼茨（1646—1716）指责牛顿“隐秘的吸引属性”是“对真正哲学之原则的混淆”，是躲进了“古老的无知避难所”。为牛顿辩护的人声称引力吸引乃是物质的一种基本属性，但牛顿本人却想找到引力的**原因**。然而，牛顿寻求答案的方法提醒我们，他并不是一位偶然生在17世纪的“现代科学家”。牛顿或许以不同于往常的谦逊认为自己仅仅是重新发现了万有引力定律，古代人对此早已知晓。这是因为牛顿信奉古代智慧（*prisca sapientia*），文艺复兴时期的许多人文主义者都持有这种看法，认为神在太古时期揭示了一种"原初智慧"，之后随时间而逐渐败坏。牛顿试图阐释希腊神话、《圣经》段落和赫尔墨斯著作，以表明其中蕴藏着关于世界隐秘结构的思想，包括他本人的平方反比引力定律。牛顿似乎认为——并相信"古人"也认为——引力吸引源于上帝在世界中持续的直接作用。正如开普勒觉得自己揭示了上帝的几何设计一样，牛顿认为自己被赋予了使命去恢复古代的知识——不仅仅是科学知识。他年复一年地研究着神学和历史，相信基督教和所有其他知识一样会随时间而败坏，遂致力于恢复据说是"原初性的"神学，例如，这种原初的神学并不包含基督的神性。同样，牛顿之所以钻研古代年代学，部分程度上也是为了获得可靠年代以解释《圣经》关于世界末日的预言。这里我们再一次回到了自然哲学（与现代科学相比）更加宽泛和全面的看法。牛顿认为"自然哲学的任务是恢复关于整个宇宙体系的知识，包括作为造物主和永恒动因的上帝"。

第四章

月下世界

近代早期的许多自然哲学家都将目光投向了天空，但重新看待地界事物的自然哲学家更多。月下世界是地球以及土、水、气、火四元素的领域，是生灭变化的领域，是不断发生转变的动态世界。重元素（土和水）和重物自然落向宇宙的最低点，即静止的地球所处的宇宙中心。轻元素（气和火）则朝着月亮天球上升，后者是四元素的最高边界。于是，经由一种基于其重性或轻性的“自然运动”，每一种元素都在宇宙中找到了其“自然位置”。这种亚里士多德体系解释了为什么石头和雨水会下落，而烟却会上升，蜡烛的火焰总是朝上。而月上世界的情况则相反，天体是由第五元素构成的，第五元素非重非轻，既不上升也不下降，而是永远围绕地球作圆周运动。近代早期人重新考察了地球、元素以及变化和运动过程，提出了各种体系来理解事物。有些人明确表示打算取代亚里士多德主义世界观，另一些人则仅仅试图改进它，几乎没有人能够完全摆脱亚里士多德的影响。观察、实验和用新的概念来理解月下世界，这些努力并未造就唯一一种通向现代科学的世界观，而是创造出各种相互竞争的世界体系，它们在整个17世纪力争获得认可和占据主导地位。

地球

近代早期的自然哲学家认为，地球和宇宙其余部分一样，年龄只有数千年。所能找到的最古老的文本《圣经》所提供的年表将人类的谱系追溯到大约6000年前。虽然只有某些读者将《创世记》第一章解释为描述了包含创世六日的字面意义上的年表（圣奥古斯丁曾在公元5世纪拒绝接受这种对字句的拘泥），但没有人认真认为，人类出现以前的地球历史还可以往回追溯得更长。最长的估计是，创世大约发生于一万年前。这并非教条使然，而是根本没有证据让人不这样想。正是在尼尔斯·斯滕森（1638—1686，其拉丁化的名字尼古拉·斯泰诺更为人所熟知）的工作中，地质历史的观念出现了。斯泰诺出生在丹麦，他先是致力于解剖学，因其解剖技巧而闻名，并且用解剖技巧作出了唾液管（今天被称为"斯滕森氏管"）等重要发现。和当时其他许多自然哲学家一样，他游历于欧洲各大学术中心，与其他自然哲学家会面并交流新知。17世纪60年代，他在美第奇的赞助下定居于佛罗伦萨，开始对托斯卡纳山丘中的岩层——我们称之为"地层"——以及包裹于其中的贝壳感兴趣。他断言，这些岩层必定曾经是逐渐沉积下来的松软泥土，因此较低的岩层必定比较高者更为古老。他认为，某个地方的地层只要不是水平的，就必定是在硬化为石头后因某种剧变而发生破裂。这些结论并未使斯泰诺增加对地球年龄的估计，毕竟，泥土可以迅速硬化成砖块，但它们的确暗示，地球表面发生过巨大变化，而岩石中保存着这些变化的记录。

到了17世纪末，一些学者——特别是在英格兰——在斯泰诺工作的基础上编写了"地球的历史"来解释当时的地球样貌。

他们大都援引全球灾难作为因果动因，在自然哲学的观念和观察资料中插入《圣经》记述和其他历史记载。托马斯·伯内特的《地球的神圣历史》（17世纪80年代）提出了六个地质时代，这六个时代被《圣经》所述的灾难性事件打断。牛顿的伙伴埃德蒙·哈雷和威廉·惠斯顿（1667—1752）提出，主要是彗星与地球相撞缔造了地球历史，导致了地轴倾斜和诺亚洪水这样的事情。

博学多才的耶稣会士基歇尔对地球表面的变化作了第一手的研究。1638年在西西里岛时，基歇尔见证了一次强烈地震和埃特纳火山的喷发。此前火山作用从未成为研究课题，这在很大程度上是因为欧洲大陆唯一的活火山维苏威火山已经沉寂了300多年，直到1631年突然剧烈喷发。基歇尔前往那里观察了持续的喷发，为了看得更清楚，他甚至下到了活跃的火山口。他观察到火山活动既摧毁了旧山脉，也形成了新山脉，使地貌产生了显著改变。他把火山的热归因于地下的硫、沥青和硝石（其结合近似于火药）的燃烧。他注意到，火和喷发出的熔岩的量太过巨大，不可能产生于山体本身，于是推测，火山必定是地球深处大火的通风口。因此他断言，地球不可能只是“大洪水过后由黏土和泥压在一起形成的，就像是一块奶酪”，而是有一种复杂的动态内部结构。他设想地球内部充满了通道和室（图8）。一些通道将火从火热的核心（他**从未**在字面上将这个核心与地狱混为一谈）转到火山口，另一些通道则允许水通过，往往是从一个海洋流到另一个海洋。大量的水流经这些通道就产生了洋流和湍动。基歇尔收集了来源各异的材料，尤其是耶稣会传教士的报告，编写了他百科全书式的《地下世界》（1665），除其他各种内容外，其中包含了显示洋流、火山和海底通道可能位置的世界地图。

图8 基歇尔，《地下世界》（阿姆斯特丹，1665）对地球隐秘内部及其火山的理想化描绘。

基歇尔是对地球上最具戏剧性的事件进行观察，而吉尔伯特（1544—1603）则是在家中静静地做实验，以发现地球的另一种不可见特征。吉尔伯特是伊丽莎白一世的御医，他研究的是一种向来令人费解的物体——磁石。他的《论磁》（1600）一书考察了磁石的性质，详述了用它们所作的实验，并且区分了磁吸引力与摩擦后的琥珀临时具有的吸引稻草的能力。［对于后一现象，他用表示琥珀的希腊词*ēlectron*创造了“电”（electrical）这个词。］他的一些实验灵感来自马里古的皮埃尔在13世纪60年代所作的实验，但吉尔伯特的研究指向了一个新的目标。皮埃尔曾用球形磁石或天然磁石——带有天然磁性的磁铁矿——发现磁石有两极，并将其命名为北极和南极。同样用球形磁石，吉尔伯特观察到置于磁石之上的铁针会精确模仿地球上罗盘针的行为。因此他得出结论说，地球是一个巨大的磁石。它也有磁极，能像球形磁石一样吸引罗盘针。（以前有人认为罗盘指向北**天**极而非地极）。简而言之，吉尔伯特用球形磁石作为地球的**模型**，通过类比推理，将球形磁石在实验过程中的现象外推到整个地球。

吉尔伯特的目标是巩固哥白尼的学说，后者使自然位置和自然运动的整个概念变得混乱。哥白尼让地球运动起来，在远离宇宙中心的地方绕轴自转和绕太阳公转，这引出了严重的物理问题。重物为什么会落到并非处于中心的地球上？是什么东西导致地球在旋转？哥白尼学说的支持者必须找到一种新物理学，从混乱中整理出秩序。一旦声称地球有磁极，吉尔伯特便强调这些磁极定义了一个真实的**物理**轴。既然自然中的一切事物都有目的，吉尔伯特认为这个轴的目的是为地球的自转作准备。此外，地球的磁性为地球赋予了内在动力，一如磁石会使铁制物体移动。吉尔伯特所谓的地球的磁"灵魂"不仅会使罗盘针指向北方，而且会使行星绕轴转动。在此基础上，吉尔伯特提出了一种"磁哲学"，认为磁性遍布和主宰着宇宙。利用相似者互相吸引——自然魔法的"共感"——的原则，磁哲学试图表明地球上的物体被自然地引向地球，而月球上的物体被自然地引向月球，从而挽救被瓦解的"自然位置"。因此，无论地球在宇宙中处于何种位置，地球物体都将落向地球。在吉尔伯特的宇宙中，磁力既维持着月上世界也维持着月下世界的秩序，他的看法深深地影响了开普勒、牛顿等人。

地球上的运动

磁哲学试图解释物体**为什么**下落，而伽利略则试图用数学描述物体**如何**下落。他造出了斜面、摆和其他仪器来研究地界运动。他在被软禁期间写的《两门新科学》（1638）是他从16世纪90年代开始的运动研究的顶峰。伽利略发现，所有物体无论重量如何，都以相同速度下落，这与亚里士多德的说法相反。他用优雅的逻辑论证说，如果滚下斜面的球速度会增加，滚上斜面的球速度会减小，那么在水平面上——既不向上也不向下——滚动的球将会保持恒定的速度。由于地球上的"水平"面实际上是弯曲的地球表面，因此在完全抛光的地球表面上滚动的球将会永远围绕地球运转。运用此"思想实验"，伽利略既阐明了一种惯性原理（运动物体会持续运动下去，除非受到外部动因的作用），又把天界的永恒圆周运动带到了地球，这进一步削弱了月下世界与月上世界的区分。

从方法论上讲，伽利略忽略的东西与他关注的东西同样重要。他描述运动时从来也不关注**什么东西**在运动，无论是一个球、一块铁还是一头牛。简而言之，他忽略了亚里士多德物理学所强调的物体的**质**。伽利略注重的是它们的**量**，它们在数学上可抽象的性质。通过剥离物体的形状、颜色、构成等特征，伽利略对物体的行为作了理想化的数学描述。一个冷的、棕色的、用橡木制成的球，其下落与一个热的、白色的立方体锡罐的下落不会有任何不同。伽利略将两个物体都还原成了抽象的、脱离语境的、能用数学处理的东西。14世纪初，一批被称为"牛津计算者"的人已经开始把数学应用于运动。事实上，伽利略在《两门新科学》中阐述运动学时，就是从他们提出的一条定理开始的。

然而，伽利略要走得更远，他将数学抽象与实验观察紧密联系了起来。他在做无数实验时，将空气阻力和摩擦看成了可以从理想数学行为中剔除出去的“缺陷”，而这些数学行为只有在思想中才能经验到。柏拉图也许会在一定程度上同意伽利略的观点，因为柏拉图所理解的世界不完美地遵循了创世所依据的永恒的数学样式（即使亚里士多德可能反对伽利略的观点）。伽利略援引基督教的“自然之书”意象说出了一句名言：“这本书，我指的是宇宙，……是用数学语言写成的，其符号是三角形、圆以及其他几何图形，没有它们的帮助，人类连一个字也读不懂。”伽利略主张把物理世界还原为数学抽象，并最终还原为公式和算法，这一技巧对新物理学的产生发挥了关键作用，是科学革命的一个显著特征。

值得注意的是，伽利略只满足于用数学来**描述**运动，而不关心运动的**原因**。伽利略工作的这个特征与亚里士多德科学有根本不同，对于后者而言，真正的知识是关于原因的知识。伽利略使用的方法与工程师类似，他更感兴趣的是描述和利用物体的**行为**，而不是**原因**。在这一点上，伽利略得益于他所处的意大利北部的环境，那里工程学发达，有学识的工程师声名显赫（见第六章）。《两门新科学》明确表明了实用工程的重要性：书中的对话者见到了威尼斯造船厂的工程，并且讨论了梁的抗拉强度及其按比例的增加和减小——这些主题对于工程师和建筑师至关重要。伽利略早年在帕多瓦大学任教授时，依靠为力学和防御工事项目提供指导来补贴其微薄的大学薪水。他后来关于抛射体运动的研究表明，抛射体的路径是一条抛物线，我们往往首先认为这是他对运动物理学的贡献。这项研究所延续的是早先尼科洛·塔尔塔利亚（1499—1557）的研究。塔尔塔利亚是一位有学问的工程师，其写于1537年的《新科学》一书将数学应用于运动尤其是炮弹的运动，这一主题对于意大利一直战火不断的诸邦来说具有直接的现实意义。我们很容易使科学发展变得过于抽象和理性，而忘记了科学往往由紧迫的实际问题所驱动。

水和空气

出于工程用途，伽利略的追随者对水的研究引出了一系列重要发现。本笃会神父贝内代托·卡斯泰利（1577—1643）是伽利略的学生，也是伽利略在比萨大学数学教席的继任者。卡斯泰利致力于研究水力学和流体动力学，这些都是重要的实际问题，因为当时的意大利建有各种供水系统，包括运河、喷泉、灌溉、沟渠和下水道等等。由于需要把水从深处（例如从深井或矿山中）竖直抽出，人们发现虹吸管无法把水向上吸到大约34英尺以上的高度。17世纪40年代初，卡斯泰利在罗马大学的同事加斯帕罗·贝尔蒂（约1600—1643）尝试用实验来研究这个问题。在基歇尔等人的合作下，贝尔蒂拿了一根能够两端封闭的36英尺长的管子，将其下端竖直插入一盆水中（图9左）。他封闭了底阀，往管内灌满水，然后封闭顶部，打开底部。水开始流出，但是当管中水柱高度下降到34英尺时，水突然不流了。是什么使水悬浮在不高不低正好34英尺处呢？

卡斯泰利的学生埃万杰利斯塔·托里拆利（1608—1647）后来继承了伽利略所担任的美第奇宫廷数学家和哲学家一职，他设计了一种简单仪器，该仪器与贝尔蒂的管子类似，但更易操作。托里拆利拿了一根长约一码的玻璃管，将其一端密封，往其中灌满水银。把它的开口端倒着插入一盆水银（图9右），此时

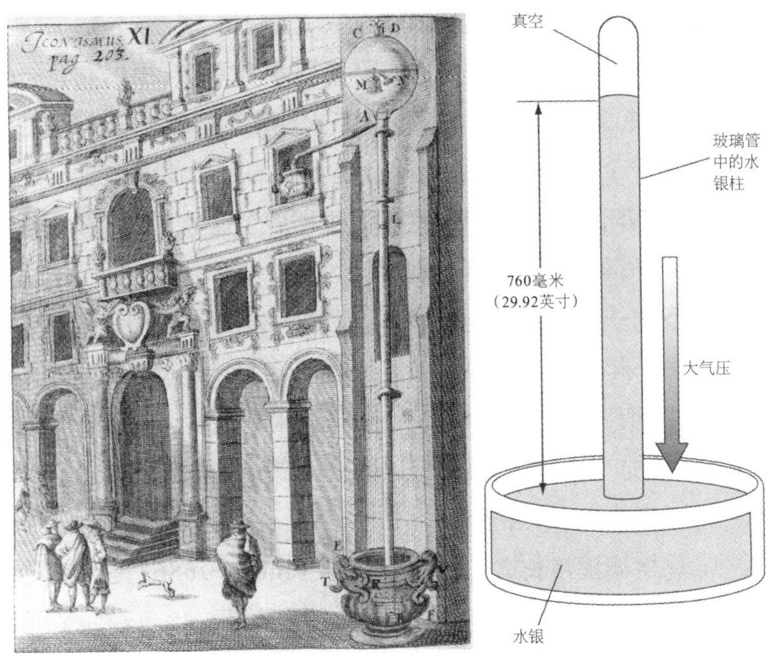

图9 （左）贝尔蒂的水气压计。加斯帕·肖特，《工艺志》（纽伦堡，1664）；（右）托里拆利简化的水银气压计的示意图。

管中水银开始排出，当管内水银柱的高度约为30英寸时停止下降，这一高度约为水在贝尔蒂管中停留高度的1/14。值得注意的是，水银的密度约为水的14倍，这意味着悬在管中的任何液体的高度都直接取决于该液体的密度。利用早先对水的研究中所得出的液体平衡思想，托里拆利解释了这些结果，声称留在管中的液体重量被向下挤压着盆中液体的外部空气的重量所平衡。空气有重量这一想法与亚里士多德体系相冲突，因为亚里士多德认为空气没有重量。托里拆利不仅提出，我们“生活在空气元素之浩瀚海洋的底部”，而且提出他的仪器可以测量和监控空气重量的变化，这使他的仪器有了一个新的名称——**气压计**

（barometer），其字面意思是“重量测量仪”。

17世纪的一些最著名的实验都是为了探究托里拆利管所激起的想法。数学家和神学家布莱斯·帕斯卡（1623—1662）提出可以用一个精巧的实验来证明，是大气的重量使液体悬浮在管中；1647年，他的姐夫弗洛兰·佩里耶做了这个实验。依照帕斯卡的指导，佩里耶在多姆山（位于法国中部，距离他们家不远）山脚下的一个修道院花园准备了几根“托里拆利管”，然后带着一根管子爬到了山上3000多英尺高的地方，发现水银面降低了三英寸。而下山之后，水银又恢复了原来的高度。在海拔较高之处，下压的“空气的海洋”较少，挤压水银的空气重量减小，因此所能平衡的管中水银也较少。

马格德堡市长奥托·冯·盖里克（1602—1686）是自然哲学家，制造了许多奇妙的仪器，而且热衷于演示。他当着许多观众的面做了一个壮观的实验，即著名的“马格德堡半球”实验。他做了两个半球形的铜壳，其边缘可以严丝合缝地合在一起。他把两个铜壳合成一个球，然后打开安装在一个半球上的阀门，用他自己仿照水泵发明的一种设备将空气从球中抽出。接着关闭阀门，冯·盖里克表明连马队都无法将两个半球分开，因为空气重量将它们保持在了一起（图10）。最后打开阀门，空气涌入，冯·盖里克轻而易举就把两个半球掰开了。

但水银上方的空间或冯·盖里克的球体之中是什么呢？许多实验者认为其中什么都没有，是**完完全全的**真空——这是一个在17世纪极具争议的话题。亚里士多德主义者和其他一些人声称真空是不可能的，他们的口号“自然憎恶真空”就反映了这一点。他们相信世界被物质完全充满，是一种**实满**，这似乎得到了一些自然现象的证实。他们认为，水银柱上方的空间

图10 冯·盖里克引人注目地表明，连马队都无法将一个抽出空气的中空球体拉开，这证明了大气的压力。加斯帕·肖特，《工艺志》（纽伦堡，1664）。

中包含着空气或者从水银中挥发出来的某种更精细的蒸汽。有一些实验试图解决这个疑问，但并没有完全解决“真空论者”与“充实论者”之间的争论。声音无法传过空间，这表明传播声音所需的空气被移走了。但光可以传过去——光难道不是和声音一样，需要某种介质来传播吗？对于当时的人来说，科学史上经常被视为“里程碑”的实验很少拥有对现代人那样的说服力。实验，尤其是对结果的解释，是一件复杂而可能引起争议的事情，过去是如此，将来也是如此。

罗伯特·玻意耳（1627—1691）很快就加入了研究空气的行列。他是英国最富有的人的幼子，因此有充足的时间和资源去献身于实验，实验地主要是他姐姐在伦敦帕尔马尔街的寓所，他成年后主要在那里生活。玻意耳和他的许多同代人指出空气可以压缩，特别是，一定量的空气所受压力越大，体积就越小，这一关系后来被称为“玻意耳定律”，今天的化学课仍然会讲授它。1658年，玻意耳听说了冯·盖里克的空气泵，于是和天才的罗伯特·胡克制造了一个改良的空气泵。它能够将一个大玻璃球抽空，使人看到密封于其中的物体在空气抽出时发生的变化（图11）。

玻意耳将一个气压计（他可能是为托里拆利管而创造了这个词）密封在他的空气泵中，看到水银面随着空气被抽出而下降。他在空气泵中做了令人眼花缭乱的实验：从试图点燃火药、用手枪射击、听手表滴答作响，一直到测量猫、鼠、鸟、蛙、蜜蜂、毛虫等各种生物在没有空气的情况下能活多久。他还用燃烧的蜡烛在空气泵中做实验，指出火依赖于空气的量。

火：化学家的工具

近代早期之前，早已有人对火作为一种物质元素的地位提出了异议。在参与这些争论的人当中，炼金术士常把火用作主要工具来研究和控制物质及其转变。科学革命是炼金术的黄金时代。今天，“炼金术”往往意味着一意孤行地（徒劳地）制备黄金，这或多或少有些“魔法”意味，从而有别于化学。但在近代早期，“炼金术”和“化学”指的是一些同样的追求。今天的一些历史学家用古体拼写chymistry来指所有这些未分化的追求。制金是化学的一个关键要素，但并不涉及（现代意义上的）“魔法”，而是一种理论基础与我们不同的活动。流传下来的一些笔记记录了“炼金术士”的日常操作，往往显示出关于实验操作、文本解释、观察和理论表述方面的细致方法论。除了追求黄金，化学还包括更广泛地研究物质，生产药物、染料、颜料、玻璃、盐、香水和油等商品。物质生产与自然哲学思辨的结合是这门学

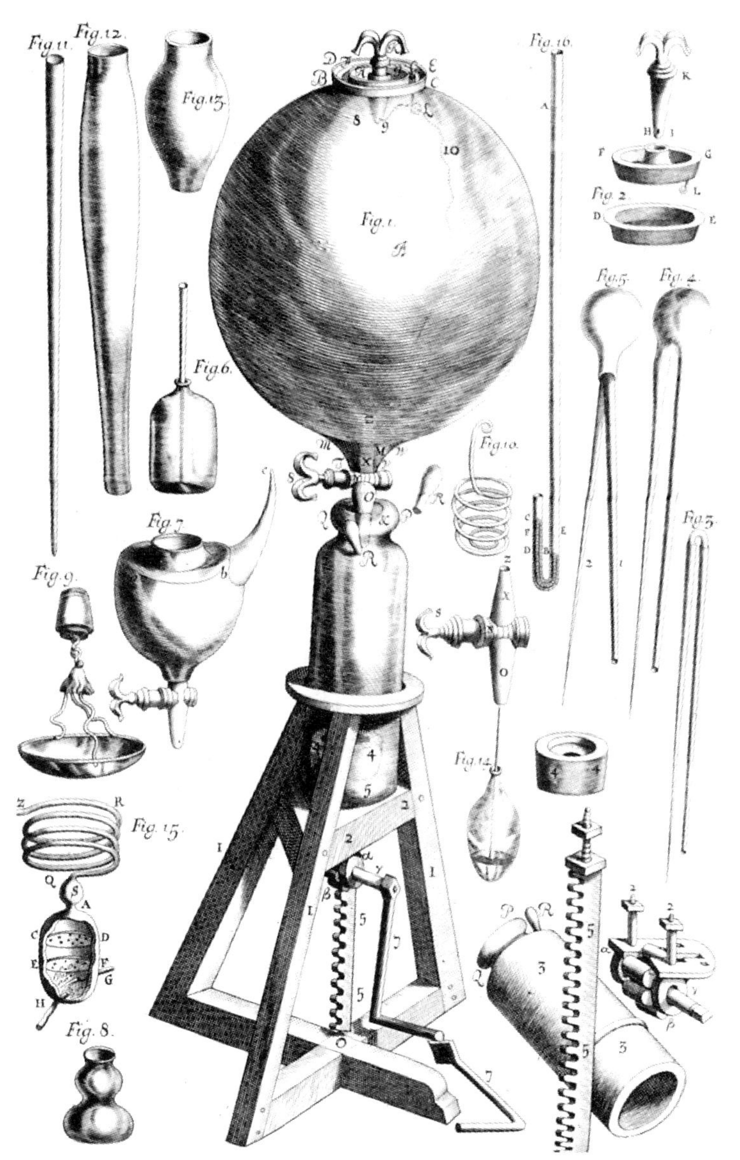

图11 玻意耳和胡克的空气泵。罗伯特·玻意耳,《关于空气弹性的物理——力学新实验》(牛津, 1660)。

科的一个核心特征，它于公元4世纪起源于希腊化时期的埃及，一直演变成今天的化学。

寻找一种方法把铅变成金并非只是一厢情愿。它基于一种理论，即认为金属是由“汞”和“硫”两种成分在地下生成的复合物。两者以正确的比例和纯度结合时就会形成金。如果没有足够的硫，就会产生银。过多的硫（一种干的易燃本原）会产生铁或铜，这表现于这些金属的易燃、坚硬和难熔。过多的汞（一种液体本原）会产生锡或铅——柔软易熔的金属。因此从理论上讲，嬗变不过是对两种成分的比例进行调整，使之符合它们在金中的比例。人们观察到，银矿石中含有一些金，铅矿石中含有一些银，这暗示嬗变是在地下自然发生的，成分不佳的复合金属被净化或“催熟”为更加稳定、成分更好的金属。问题是如何通过人工手段更快地实现这种转变。于是，制金者试图制备他们所谓的“哲人石”，这是一种引发嬗变的物质动因。据说一旦在实验室制备出来，少量哲人石几分钟之内就能把与之混合的熔化的贱金属变成金。许多文本都声称成功地实现了这个过程，追求嬗变的人力图对此进行重现。困难在于这些著作会故意保密——它的成分、过程甚至是理论都掩藏在暗号、封面名称、隐喻和往往怪异的图案标志背后（图12）。

炼金术的保密性部分源于手工活动，因为有必要将专利权保持为行业秘密。出于对货币贬值的恐惧，中世纪的法律禁止嬗变，这进一步加强了保密性。此外作者们还声称，他们的知识倘若落入坏人之手会很危险，而且这些知识是一种优越的知识，不能泄露给配不上它的人，因此需要保密。

英国人一直用“化学家”意指“药剂师”，这起源于近代早期，那时大多数化学家都至少把部分精力投入制药。把化学用

图12 一则炼金术讽喻，描绘了金和银的提纯，这是制备哲人石的第一步。国王代表黄金；跳过坩埚（一种用来提纯金属的器皿）的狼代表辉锑矿，这种材料能与银和铜起反应，除去常与金混合的银和铜；女王代表银；老人代表铅，指用铅来提纯银的灰吹过程。《赫尔墨斯博物馆》（法兰克福，1678）。

于医药始于普罗旺斯的方济各会修士鲁庇西萨的约翰（1310—约1362），他提倡用从酒中蒸馏出来的酒精来制备药用提取物。把化学用于制药在整个15世纪迅速发展，最终在帕拉塞尔苏斯（1493—1541）这位具有传奇色彩的人物那里得到了最热情的拥护。帕拉塞尔苏斯批判了以希腊、罗马和阿拉伯的作者们为基础的传统医学，并且基于从直接观察到日耳曼民间信仰的各种来源提出了自己的体系。他提倡以化学为手段把几乎所有物质都转化成一种强大的药物，对制金他并没有什么兴趣。他的指导思想是，有害性质起因于原本健康的物质中的杂质，这就像罪恶和死亡污染了那个由上帝所造的、本质上美好的世界。化学利用蒸馏、发酵和其他实验室操作，提供了区分好坏、辨别良药与毒药的方法。帕拉塞尔苏斯还告诉我们，所有物质都是由汞、硫、盐这三种主要成分构成的，这是地界的三位一体，被称为“三要素”（*tria prima*），反映了神的三位一体和人的三位一体本性——肉体、灵魂和精神。帕拉塞尔苏斯所谓的“炼金术”（*spagyria*）过程力图将一种物质分成其三要素，并分别进行提纯，然后将其重新组合成一种“高贵”的原始物质，它的药力更强而且没有毒性。

但帕拉塞尔苏斯更进了一步：化学不仅是一种用来制药的工具，而且也是理解宇宙的关键。帕拉塞尔苏斯在16世纪末的追随者对其往往混乱的著作（有人传言他是酒醉时口授的）做了系统整理，提出了一种化学论（chymical）世界观，把几乎一切事物都设想成本质上化学的。雨水经过海洋、空中和陆地而完成的循环是一次大蒸馏。地下矿物的形成，植物的生长，生物的产生，消化、营养、呼吸和排泄等身体机能，这些都被视为本质上化学的。上帝本身没有变成柏拉图主义者所说的几何学家，而是变成了化学大师。上帝从原始的混沌中创造出一个有序的世界，就类似于化学家将普通材料萃取、提纯和转化为化学产品；上帝用火对世界进行末日审判，就类似于化学家用火把杂质从贵金属中清除出去。这种世界观甚至把人的最终命运看成化学的。人死后，灵魂和精神脱离肉体。物质性的肉体在坟墓中腐烂，直到全部死者复活时得到更新和转变，作为化学家的上帝将净化后的灵魂和精神重新注入其中，产生出一个荣耀的、永生的人，一如在帕拉塞尔苏斯所说的炼金术中，三要素从物质中分离，提纯后重新结合成一种“光彩夺目的”产物。

帕拉塞尔苏斯的学说吸引了众多追随者。1572年第一次看到新星的不久前，第谷还在实验室中根据帕拉塞尔苏斯的学说制备药品。后来，第谷在他的天文台城堡中建了一个实验室，旨在研究他所谓的“地界天文学”，即化学（所谓“上行下效”）。帕拉塞尔苏斯的风格具有反体制性（往往表达为痛骂古典学问、大学和执业医师），因此他的观点引发了激烈的争论，其追随者往往集中于体制外。事实上，整个化学在大多数时间都存在于传统的学术机构之外，地位很是尴尬。物理学和天文学从中世纪以降就是大学必须研究的科目，但化学直到18世纪才获得了学术地位。其中一个原因是，它无法夸耀其古典根源，无论是亚里士多德还是任何其他古代权威都没有写过化学论著，这一点不同于天文学、物理学、医学和生命科学。化学与商业和人工制品有密切关联，具有实用性，而且往往混乱不堪、艰苦费力、气味难闻，这些都使化学无法令人尊敬。然而，注重实用性的实验也意味着化学能够收集大量材料，认识它们的属性，掌握处理它们的能力。在整个17世纪，这些知识在商业上的重要性与日俱增，许多化学家因此走上了一条实业道路——有时是被王侯或其他主顾以及采矿业所雇佣，旨在提高产量或探索物质的嬗变；有时是独立工作，旨在为市场推出新的商品。不幸的是，化学仿造宝石和金属的能力以及关于制造黄金的说法为诈骗提供了可乘之机，导致人们普遍将化学与不道德的勾当联系起来。早在中世纪晚期，但丁就把化学家——“自然的模仿者”——与伪造者一道置于地狱的第八圈，后来，17世纪剧作家本·琼森在其《炼金术士》（1610）中也用弄虚作假的化学家及其贪婪的客户来增强喜剧效果。

17世纪的化学训练大都是在医学环境中进行的。在德国，约翰内斯·哈特曼（1568—1631）于1609年成为第一位**化学医学**（*chemiatria*）教授。授予他此项教职的马堡大学是黑森—卡塞尔公国的莫里茨王子新建（因此更能创新）的一所加尔文主义机构，莫里茨王子的宫廷资助了许多制金者、帕拉塞尔苏斯主义者和其他化学家。在法国，常规的化学教育开始于巴黎的国王花园（Jardin du Roi），这是一个旨在传播和研究药用植物的植物园。一系列讲师在花园基于面向公众的实验演示来讲授实用课程。私人讲师往往是药剂师，他们也讲化学课，比如尼古拉·莱默里就在其巴黎寓所内讲课，他的教科书《化学教程》（1675）成为畅销书。事实上，在法国和德国出版的数十本化学教科书确立了一种教学传统，弥补了化学在大学课程中的缺失。

化学的实用色彩并不意味着它没有对自然哲学理论作出重要贡献，事实恰恰相反。17世纪最重要的进展之一，即原子论的复兴，便是部分建基于化学观念和化学观察。13世纪末的一位被称为盖伯的拉丁炼金术士，已经用一种准微粒的物质理论来解释化学性质。例如，他设想金的“最微小部分”被紧紧挤在一起，其间不留空隙，从而解释了金的密度和耐腐蚀性。而铁的“最微小部分”排列得更为松散，留下的空隙使铁的重量更轻，并且为火和腐蚀剂进入铁、将其分解为铁锈提供了空间。后来的化学家继续发展稳定的微粒这一思想，并借此来解释他们观察到的现象。主流的亚里士多德主义者常常会拒斥这样的观念，因为他们声称，物质在结合后会失去自己的身份。但从事实际研究的化学家知道，他们往往可以在一系列转化的结尾恢复初始材料。例如化学家知道，用酸处理的银“消解”为一种清澈的同质液体，它可以自由地穿过滤纸。用盐处理时，该液体会析出一种重的白色粉末。如果把这种粉末与炭相混合，并且将其加热至赤热状态，就会重新获得原有重量的银。这个著名的实验表明，银自始至终都保持着自己的身份，尽管从外表上看似乎并非如此，尽管它被分解成了能够透过纸孔的看不见的小微粒。化学操作为这些“原子”提供了最好证据。

原子论和机械论

微粒物质观的化学传统与古代原子论的复兴相互交织。古希腊原子论始于公元前5世纪的留基伯和德谟克利特。他们设想了一个由不可分割的原子所构成的物质世界，原子在虚空中不断分散和聚集，其千变万化的组合导致了我们看到的所有变化。他们的思想大部分已经湮没于古代历史中。亚里士多德对其作了详细反驳。虽然伊壁鸠鲁（前341—前270）把原子论当作自己的道德哲学的基础，但是当伊壁鸠鲁主义因其无神论和享乐主义倾向（伊壁鸠鲁并无这两种意图）而不再受到青睐时，原子论一同遭到了抛弃。直到罗马人卢克莱修普及伊壁鸠鲁思想的长诗《物性论》于1417年被重新发现，原子论才得以复兴。但卢克莱修对原子论与无神论之间关联的强调，使他的著作最初无法得到认同。具有讽刺意味的是，伊壁鸠鲁原子论得以恢复名誉要归因于一位神父——伽桑狄（1592—1655）。伽桑狄否认原子是永恒的（只有上帝是永恒的）和自行移动的（是上帝使它们移动），主张人的灵魂是非物质的和不朽的，并且建立了一个全面的世界体系，把不可见的微粒及其运动用作基本的解释原则。他的体系以及其他类似的体系后来被称为“机械论哲学”。

机械论哲学认为，所有可感的性质和现象都源于不可见的物质微粒（有时也被称为原子、微粒，或径直称为粒子）的大小、形状和运动。严格的机械论哲学家坚持认为，万物都是由同一种“原料”构成的，我们所觉察到的各种物质和属性都是源于这种原料微粒的不同形状、大小和运动。与其重数量、轻性质的态度相一致，伽利略认为冷、热、颜色、气味和味道等大多数性质实际上并不存在，而只是微粒作用于我们感官的结果。对于伽利略以及后来的机械论哲学家而言，唯一真实的性质——**第一性质**——是微粒的大小、形状和流动性。所有其他性质都是**第二**性质，它们只存在于感知者中，而不存在于被感知者中。在机械论哲学家看来，醋之所以显得酸，仅仅是因为尖且锐利的醋微粒刺痛了舌头。如果没有舌头，“酸”就没有任何意义。玫瑰显示为红色，仅仅是因为玫瑰的微粒以特定的方式改变了反射光，而改变后的光又作用于我们的眼睛。玫瑰之所以芳香，是因为玫瑰花散发出的微粒经由空气进入了我们的鼻子，在那里撞击嗅觉器官，由此产生的运动被输送到大脑，并被转换成一种气味的感觉。这种观点从根本上反对亚里士多德看待世界的方式，在亚里士多德的世界观中，可感性质实际存在于物体之中，对于解释物体的性质和效应起着至关重要的作用。

该体系在两种意义上是机械论的。首先，结果都是通过机械接触引起的，比如锤子砸到石头上，或者弹子球相互碰撞。超距作用或共感的力量在其中没有地位。其次，世界以及其中的物体——甚至是具有广泛影响的笛卡尔机械论哲学中的植物和动物——都被理解成**机器**。机械论哲学家把世界比作一个复杂的钟表装置，就像当时巨大的机械钟，隐藏于其中的齿轮、重锤、滑轮和杠杆使外面的表针转动，钟鸣响，小铸像翩翩起舞、鞠躬致意，机械公鸡喔喔啼叫，一切都符合完美的秩序和规则性。“世界机器”（*machina mundi*）一词的历史可以追溯到卢克莱修，一些中世纪作者用它来表达宇宙的复杂规律性，但对于那些作者而言，**机器**的意思更像是框架或结构，表达的是天地万物各个部分的相互依存关系。而机械论哲学家却为这一图像赋予了一种自动机的含义，即某种人工的东西，但却机械地模仿一个生命体的活动。机械论观点反映了当时日益增长的技术能力，对世界的理解渐渐从活的生物模型变成了无生命的机器。这种观点甚至导致了对上帝本身的重新理解。上帝不再是一个几何学家、化学家或建筑师，而是越来越被看成一位机械师或钟表匠，一个对世界机器进行设计和组装的技师。这一形象在17世纪末的英格兰变得尤为根深蒂固，它构成了关于“智能设计”的现代讨论的基本背景。在近代早期，随着神学和自然哲学的彼此融合，科学概念和宗教概念一同发展和成熟起来，彼此发生影响，互相作出回应。

机械论哲学家力图用他们的原则来解释所有自然现象，一个棘手的问题是如何解释“隐秘性质”、共感和超距作用，它们曾使亚里士多德主义者感到沮丧，也是自然魔法的基础。机械论者所青睐的解决方案是诉诸一种不可见的物质流溢——是微粒“流”将影响从一个物体带到了另一个物体。例如，火之所以能够加热远距离处的物体，是因为快速移动的火微粒从火焰中散发出来并且击中了物体。其他解释则需要有更具创造性的解决方案。笛卡尔对磁吸引力的解释是，磁体发射出一种恒定的螺旋形微粒流。他假定铁含有螺旋形的孔洞，磁体发射出的微粒进入了铁的孔洞，在其中旋转，从而把铁“拧”得更靠近磁体。即使是看到血腥场面时不自觉地转头这一反射动作，也要通过尖的微粒流会伤害眼睛来解释。

玻意耳不仅提出了“机械论哲学”一词，而且特别把它与化学结合了起来，因为他认识到化学在揭示世界运作方面具有特殊能力。玻意耳的研究涉及17世纪化学的所有四个主要方面：制金、医药、商业和自然哲学。他热切地寻求着研制哲人石的秘密，并试图接触可能提供帮助的“秘密行家”。他声称目睹过哲人石的使用，验证过由铅生产出的金，并促使一部禁止金属嬗变的英国法律于1689年被废止。他收集了新的化学药品，特别是那些不太昂贵的、救济穷人的药品（和今天一样，那时的医疗护理和药品的价格也过高）。他还主张把化学用于实用目的，改进行业、贸易和制造。也许最有名的是，他宣扬化学是研究世界的最佳途径，并且努力提升化学的地位。玻意耳解释说，他之所以投身于被他的朋友们视为“一种空洞的欺骗性研究”的化学，是因为它为机械论哲学家所提出的微粒体系提供了最好的证据。例如他用实验表明，硝石可以产生一种不变的碱性盐和一种挥发性的酸性液体，两者结合会重新产生硝石。他的结论是，复合物质可以分成小块，这些小块一起还原会重新形成原先的物质，就像一台机器的零件。虽然玻意耳拒绝接受帕拉塞尔苏斯的大部分学说，但（他所谓的）这种“重新合并”与“炼金术”惊人地相似，事实上，玻意耳的想法正是以前面提到的那种历史悠久的制金和化学医学两种传统为基础的。

机械论哲学在17世纪末逐渐衰落。玻意耳变得不那么热衷于它，因为他意识到这一哲学的过度扩展可能会导致决定论、唯物论和无神论。假如世界仅仅是一些相互碰撞的微粒，那么自由意志或神意将没有位置。如果上帝是一个钟表匠，那么他是先开动世界然后对其不闻不问，还是如同一个拙劣的机械师，必须经常对其重新调整？化学家一直对严格的机械论哲学兴趣不大，因为他们平日里看到的大量属性似乎无法通过同一种物质不同形状的微粒这般贫乏的观念来解释。同样，生命过程过于复杂，超出一定限度便无法用简单的力学来解释。最后，牛顿所说的吸引力是一种超距作用，无法对其进行机械论解释。牛顿主义的胜利其实意味着严格机械论的失败。

不断演进的亚里士多德主义

亚里士多德和亚里士多德主义在本章已经多次出现。事实上，有一种关于科学革命的解释是，科学革命完全是对一种垂死的经院亚里士多德主义的拒斥。但这种观点没能认识到经院哲学的弹性和不断演进。17世纪各种“新”哲学的支持者经常用严厉的措辞讽刺和批判亚里士多德主义，但其他自然哲学家始终处于“亚里士多德主义”框架内，继续更新着系统，做着卓有成效的工作。无论在中世纪晚期还是在近代早期，“亚里士多德主义”或“经院哲学”都不意味着顽固地秉持亚里士多德本人所做的任何断言。即使是亚里士多德最伟大的学生特奥弗拉斯特，也是通过在一些观点上不同意亚里士多德的看法而延续着亚里士多德传统。在中世纪，自然哲学家普遍引述亚里士多德，但经常只是作为自己研究的一个出发点，所得出的结论往往与亚里士多德的结论相反。到了文艺复兴时期，存在着许多不同的甚至是相互冲突的亚里士多德主义。

自然哲学的实验进路和数学进路并非亚里士多德本人工作的关键部分，但对于17世纪的亚里士多德主义者来说，它们变得越来越重要。耶稣会士是明确致力于坚持一种亚里士多德主义自然哲学的最明显例子，但里乔利和格里马尔迪等许多耶稣会士都做了与伽利略运动学有关的大量实验，而且把一些明显与亚里士多德相矛盾的观念和发现包括进来。同样，耶稣会士尼科洛·卡贝奥（1586—1650）拒绝接受吉尔伯特关于其磁学实验支持了哥白尼这一解释，但卡贝奥自己也做了大量磁学实验。到了17世纪末，耶稣会士已经在一种“亚里士多德主义”框架中采用了伽桑狄和笛卡尔所阐述的许多机械论观点。在其支持者看来，经院哲学仍然是一种有用而灵活的自然研究**方法**。他们虽然对17世纪的许多创新持保守态度，但仍然是科学革命的积极参与者和贡献者。

在科学革命过程中，亚里士多德主义的确遭遇了截然不同的强劲的竞争对手，这是在中世纪晚期没有遇到过的。在整个近代早期，新的世界观——磁的、化学论的、数学的、自然魔法的、机械论的，等等——均作为挑战者和貌似合理的替代品而出现，而经院哲学则力图将新的材料和观念吸收到一种“亚里士多德主义”框架中。不同世界体系的捍卫者之间的持续争论不仅引出了各种论战技巧，而且引出了对如何建立一种新的、最好是全面的自然哲学这一紧迫挑战的大量不拘一格的回应。从我们现代的角度来看，很难想象近代早期会发展出如此众多关于基本问题和方法的不同观点和进路，也很难想象越来越多的自然哲学家会以如此的热忱富有成果地探索他们的世界，并且设计出大大小小的体系来尝试理解所有这一切。16、17世纪之所以的确是“革命性的”，这是一个重要原因。

第五章

小宇宙和生命世界

除了月上世界和月下世界，还有第三个世界引起了近代思想家的注意，那就是人体这个“**小宇宙**”或“小世界”。近代早期的医生、解剖学家、化学家、机械论者等等都密切关注我们所寄身的这个生命世界。他们探索其隐秘的结构，力图理解其功能，希望找到新的健康之道。人体的生命活力自然将人与地球上的其他生命——植物和动物——联系起来。在科学革命时期，这些生物的名录在激增，这不仅要归因于探险航行，也是因为显微镜被发明出来。显微镜揭示了平凡之物背后人们从未想到的复杂世界以及一滴水中所包含的新生命世界。

医学

人体是医生的首要关注对象，在整个中世纪盛期和近代早期，医学一直具有很高的社会地位和学术威望。医学院连同法学院和神学院构成了大学中的三个高等学院。1500年左右大学所传授的医学知识，是中世纪的阿拉伯人和拉丁人经验的累积，和以古希腊罗马的医学学说为基础所作的创新。盖伦、希波克拉底、伊本·西纳（又称阿维森纳，约980—1037）是主要权威，体液理论是这些医学知识的基础。体液理论主张，身体健康不仅需要各种器官正常运作，还需要体内四种体液的平衡。它们分别是血液、黏液、黄胆汁和黑胆汁，其相对比例决定了人的**气质**。四体液与亚里士多德的四元素相对应，遵循着后者的原初性质配对（图13）。

医生的职能是帮助重新建立体液平衡，这可以通过规定特种饮食、日常养生法和药物来实现。这种以盖伦学说为主导的医学使用的是“对抗疗法”，也就是说，如果病人由于体内黏液（冷和湿的体液）过多而引发了“感冒”（cold，我们今天仍然沿用了盖伦医学对它的称谓），那么他可以服用热和干的食物和药物

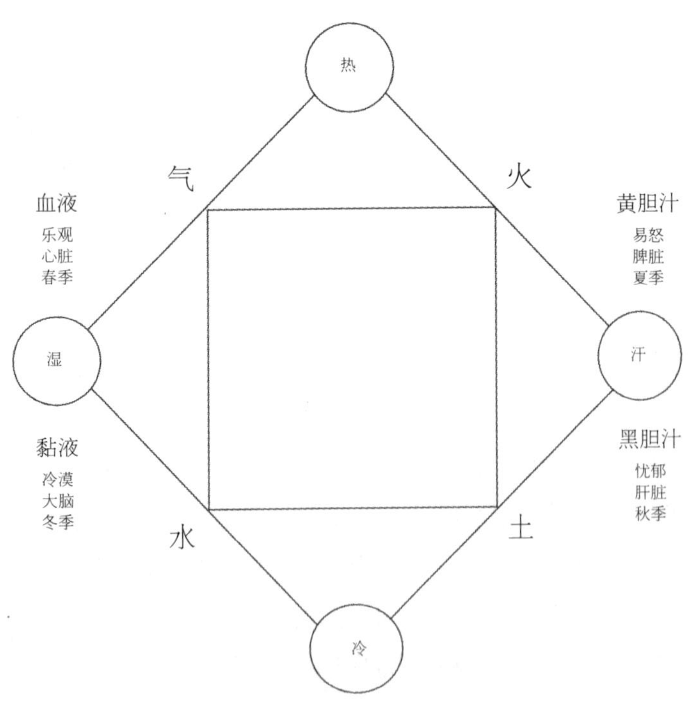

图13 “元素正方形”，显示了四元素的性质及其与四体液、四体质和四季的关系。

来帮助恢复体液平衡。发烧的病人则需要冷和湿的药物，洗冷水澡，或者通过放血疗法来排出多余的血液及其热性。

人们认为，月上世界与人体之间存在着许多关系，这充分表明了近代早期世界的关联性。大宇宙对小宇宙的影响基本上未被质疑，尽管这种相互作用的细节始终是有争议的。于是，占星学在诊断和治疗中发挥了关键作用，占星学的主要应用可能是医学而不是预测。每一个身体器官都对应着黄道上的一个宫，特别容易受到那颗在性质上与之相似的行星的影响（图14）。

例如，大脑是一个冷而湿的器官，它受月亮——一颗冷而湿的行星——的影响最大。[因此，大脑失常的人今天依然被称为lunatic（精神失常）——来自拉丁词luna（月亮）——或者更通俗地说是moony（发狂的）]。了解疾病发作时的行星位置有助于医生认识主要的环境影响，或者确认受到潜在影响的身体器官，从而作出诊断。不仅如此，每个人的体液都有一种独特的比例——被称为他的“**体质**”，这取决于人出生时占主导地位的行星影响。这意味着每个病人都必须恢复其特有的体质。

近代早期医学中没有整齐划一的治疗。治疗必须因人而异，同一种药物不会适用于所有人，特定的饮食和养生法也要与治疗并行。因此，为了弄清楚病人的体质，医生也许会考察病人的命盘。依据希波克拉底提出的“危险期”思想，在疾病的发展过程中存在着一些“危机”点，病人要想康复必须成功战胜它们，因而占星学计算可能同样有助于安排治疗时间。诊断也依赖于尿液检查，便携的参考图列出了一览表，显示了病人尿液的颜色、气味、浓度甚至是味道以及这些指标与不同疾病的关系。脉搏的快慢、节奏和强度也是如此。

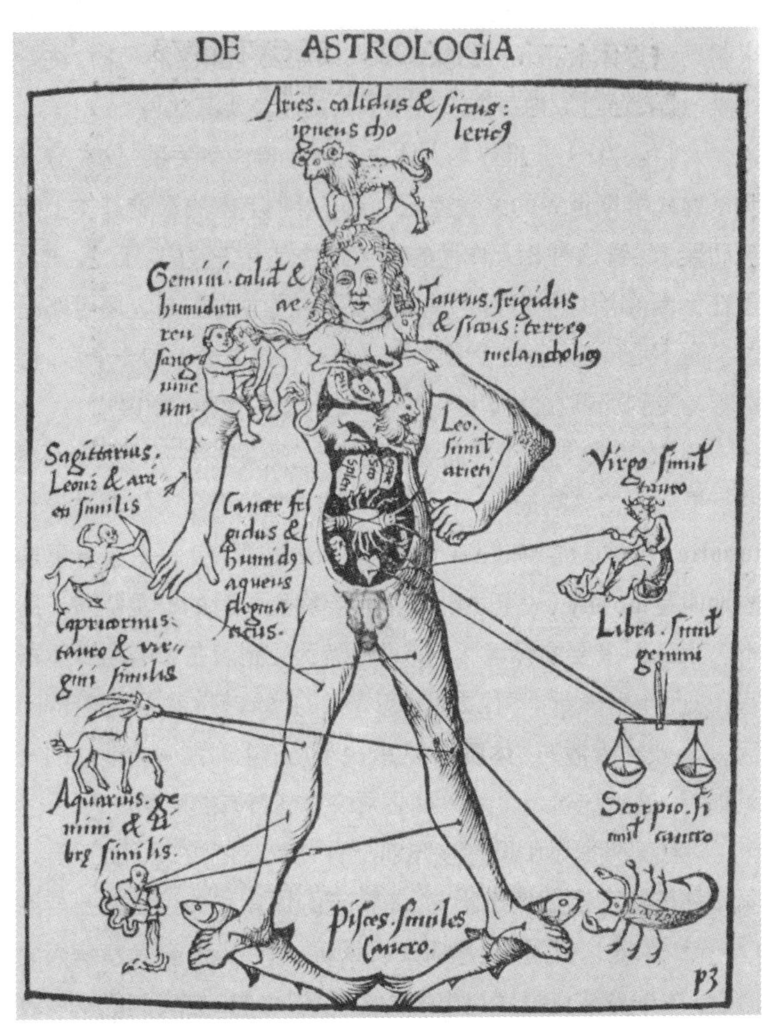

图14　该图显示了人体器官及其与黄道各宫的对应。源自近代早期的一部百科全书，格雷戈尔·赖因施编，《哲学珠玑》（弗赖堡，1503）。

在科学革命时期，医疗方法至少在执业医师那里并未发生显著改变。虽然在新的观念和发现的影响下有缓慢发展，但以盖伦和希波克拉底为核心的医学传统一直延续到18世纪（尽管占星学诊断在17世纪就已开始衰落)。这一延续既反映出医学院课程的稳定性,也反映出了医学行会或医学职业认证机构的保守态度。于是,创新往往来自于非执业医师。然而,严格的医生职业许可仅限于那些大的城市中心。在大多数地方,维护人们健康的都是未受过或只受过些许正规医学教育的各类治疗师,其数量远远超过执业医师。几乎家家都为自己或邻居备着一系列家用医疗药品。药剂师使单方药和成药变得唾手可得,以至于几乎所有人都可以合成奇异药物(此举有时很危险)。手术一般由外科理发师(barber-surgeon)来做,与一般的外科医生相比,他们的地位较低,所受教育也不够正规。"江湖医生"(即提供各类药物和治疗的无照医生)发现还是城市中的生意最好做,虽然伦敦、巴黎和其他大城市经常禁止他们行医。与现代医疗迥异的是,有些治疗需要签订合同,换言之,医生的报酬依赖于治疗效果。

非执业医师对全新的医学研究方法(比如帕拉塞尔苏斯理论和17世纪的海尔蒙特理论)表现得更加热情,这往往对现有的医学体系构成了直接挑战。不过,医学中新的化学方法缓慢而持续地进入了官方药典和专业机构(如1518年成立的伦敦皇家医师学院)的医疗活动中。在法国,保守的巴黎医学院支持盖伦的医学,而蒙彼利埃医学院则支持帕拉塞尔苏斯的医学,两者针对化学药物的风险和回报展开了长达数十年的争论。这一冲突也反映了皇家的、集权的和占主导地位的巴黎天主教徒与外省的、新教徒为主的蒙彼利埃人之间的分歧。他们最激烈的争论涉及锑这种有毒矿物的医学用途,这场争论直到1658年才结束。当时路易十四在一次战役中病倒了,皇家医生的传统治疗不起作用,当地医生用含锑的酒使路易十四呕吐而治好了他的病。

此后巴黎医学院不得不投票肯定了帕拉塞尔苏斯派使用“催吐药酒”的正当性。

解剖学

解剖学在近代早期取得了重大进展。尽管盖伦强调解剖学在古代的重要性，但罗马人认为解剖对尸体造成的冒犯在社会和道德上都是不可接受的，于是盖伦只能对猿和狗进行解剖，并将其发现类比于人体。（不过在担任角斗士医生期间，他无疑经常能够看到暴露的人体内脏。）在古代，只有埃及允许人体解剖，这可能是因为制作木乃伊需要经常开膛破肚、切除器官。然而在中世纪晚期，人体解剖在帕多瓦和博洛尼亚等意大利大学的医学院已被普遍接受。到了1300年左右，作为教学的一部分，医科学生必须观察人体解剖。认为天主教会禁止人体解剖不过是19世纪的一则毫无依据的谎言。当时人体解剖的主要限制因素是尸体短缺。由于体面的人不允许自己或其亲属的尸体在观众面前被陈列和切割，解剖用的尸体主要来自于死刑犯，特别是外国死刑犯。

到了16世纪初，人们对人体解剖兴趣大增，尤其是在意大利，以1543年——哥白尼的《天球运行论》也出版于这一年——安德列亚斯·维萨留斯（1514—1564）的里程碑式著作《论人体结构》的出版为顶峰。维萨留斯出生于佛兰德斯，在帕多瓦大学学习，获得硕士学位后担任外科学讲师。在一位恰当安排了死刑执行时间（由于缺少冷冻或保存措施，尸体必须马上解剖）的法官的协助下，维萨留斯进行了大量细致的解剖。他注意到了盖伦等人的错误，以新的方式对人体各个部分作了归类，不仅依据其功能也依据其结构。在维萨留斯的亲自监督下，提香工作室的艺术家们凭借高超技艺绘制了精细的解剖图。这是《论人体结构》的一个主要特色，书的正文详细解释了每一幅插图和相应的解剖学特征。倘若没有印刷术，制作插图如此丰富的书是不可能的。但这部巨著仍然过于昂贵，促使维萨留斯又制作了一部供学生使用的廉价本，他的思想、发现和组织原则也因此广泛流传开来。对解剖学与日俱增的兴趣促使人们修建了解剖室，它首先出现在帕多瓦大学（1594），接着在莱顿大学（1596）、博洛尼亚大学（1637）等地。虽然旨在服务于医科教学，这些解剖室——尤其是北欧的那些——也引起了一般公众的兴趣。

解剖并不限于人体或医学院。随着科学社团在17世纪的兴起，动物解剖成为社团活动的一个重要部分。17世纪70和80年代，成立不久的巴黎皇家科学院收到了在路易十四动物园中死去的异国动物的尸体，包括鸵鸟、狮子、变色龙、瞪羚、海狸、骆驼各一只。解剖骆驼时，科学院院长克劳德·佩罗（1613—1688）不小心被解剖刀划伤，死于感染。在17世纪50和60年代的牛津和伦敦皇家学会，一些人在实验中不仅解剖尸体，而且也对活体动物特别是狗进行解剖。许多实验在现代读者看来过于恐怖（玻意耳即因此而心绪不宁）。这些实验力图了解神经、肌腱、肺、静脉和动脉的实际运作。实验中常常会注入各种液体，观察它们在身体中的流动及其生理效应，有时会进行跨物种的输血，甚至为了治病而把健康的羊血直接输给病人。

对血液和人体内液体运动的兴趣部分源于威廉·哈维（1578—1657）1628年发表的主张血液循环的观点。根据盖伦的说法，静脉系统和动脉系统是分离的。肝脏持续产生深色的静脉血，静脉把这些营养液输送到全身。一部分静脉血流入心脏，并且流经将左右心室分隔开来的膈膜上的孔洞。经由肺动脉从肺部抽出的空气把静脉血变成了鲜红的动脉血，然后通过动脉系统为全身输送营养。没有血液返回心脏。但16世纪的解剖学家发现了盖伦体系的问题。他们质疑心脏膈膜中是否存在孔洞，并且发现肺动脉充满了血液而非空气。后一发现引出了“小循环”的猜想：静脉血由心脏出发，经由肺返回心脏，然后流到全身。在帕多瓦大学，哈维跟随当时一些最伟大的解剖学家学习，特别是吉罗拉莫·法布里齐奥·阿奎彭登特（1537—1619），后者描述了自己在静脉中发现的“瓣膜”。哈维后来说，此发现促使他开始思考一个更大的血液循环系统。

哈维指出，倘若血液没有以某种方式再循环，则心脏泵出的血液量很快就会耗尽全身的血液供应。他用绷带来选择性地阻止血液流动，用实验导出了“大循环”，即心脏通过与之相连的动脉系统和静脉系统循环地泵出血液。哈维认为血液循环运动令人满意，因为它意味着小宇宙模仿着天界大宇宙，这个大宇宙的自然圆周运动在亚里士多德看来是最完美的运动。实际上，哈维坚持着亚里士多德的进路和方法，对心脏和和血液的重视也部分缘于亚里士多德为其赋予的核心地位。这个例子再次说明，亚里士多德在科学革命时期一直很重要。然而，哈维无法找到连接静脉与动脉的毛细血管。哈维去世四年后，马切洛·马尔比基（1628—1694）才第一次发现了这些结构。马尔比基发现，流经毛细血管的血液将青蛙透明的肺部组织中的静脉系统与动脉系统连接了起来。他推测，类似的血管连接了全身的静脉和动脉。为了进行这一观察，马尔比基使用了一项新近的发明——显微镜。

显微镜、机械论和生成

关于显微镜在17世纪初的起源，我们并不很清楚，但是和它的伙伴望远镜一样，显微镜揭示了一个新世界，激发出了新的思想。伽利略曾经用一个与望远镜类似的装置把小物体放大，但第一批显微镜绘图出现在弗朗切斯科·斯泰卢蒂和费代里科·切西1625年所做的蜜蜂研究中。他们将该研究献给了教皇乌尔班八世，因为蜜蜂是教皇所属的巴尔贝里尼家族的族徽。17世纪60年代，胡克制作了一架改良的显微镜来考察各种东西，从虱子等小昆虫到霜晶，再到软木的精细结构，不一而足。他发现软木被分成了一个个腔室，并称之为“修道院单人小室”（cell），因为它们非常类似于修道院的住处。安东尼·范·列文虎克（1632—1723）是荷兰代尔夫特的一名布商和测绘员，他设计了当时最简单且最强大的放大镜。他用一个极小的玻璃珠作为单透镜，制作了500多架显微镜，发表的显微学观察之多可以说前无古人。他将各种东西置于显微镜下，观察到了人和动物精液中的“虫子”、血液中的血球（以及它们在小鳗鱼尾部毛细血管中的流动）、牙垢中的细菌，还有池水和植物浸剂中成群的“小动物”。他对精子的发现引发了关于动植物生成的本性的热烈讨论。列文虎克本人支持**预成论**，主张新生后代的微小版本存在于各自的精子中，或者根据一些同时代人的说法，存在于各自的卵中。与此相反的**渐成论**则主张，胚胎结构是在妊娠期间的各个阶段重新产生的。预成论尤其吸引机械论哲学家，因为它将生成归结为单纯的机械生长，即微小的有机体通过吸收新物质而逐渐长大。预成论由此抛弃了渐成论者大都认为不可或缺的非物质生命活力。渐成论者认为只有依靠生命活力，无形的物质——精液和/或经血或卵液——才能变成有组织的、分化的胚胎。作为渐成论者，哈维打碎了处于不同发育阶段的鸡蛋，观察到血液最先形成，他认为这证明血液是生命和一种主导幼体形成的生命灵魂（vital soul）的存在之所。然而，预成论同样引出了问题，即新有机体的微小形式位于何处以及实际上何时开始出现。少数人提出，未来所有后代都被一个套一个地包含在上帝创造的某个物种的第一个个体中。

显微镜揭示出了生命体中类似机械的结构，机械论者对此尤为兴奋，因此17世纪末的显微学家大都是机械论者。他们之所以支持哈维的血液循环理论，部分是因为它将心脏刻画成一个泵（一个机械装置，尽管哈维本人远非机械论者）；他们力图将复杂的生命系统归结为机械原则。例如，乔万尼·阿方索·博雷利（1608—1679）在佛罗伦萨用简单机械来分析动物的运动，将骨骼、肌腱和肌肉理解成杠杆、支点和绳索，将体液和血管理解成液压装置和水管，从而创立了后来所谓的生物力学。尼赫迈亚·格鲁（1641—1712）在伦敦探索了植物隐秘的解剖结构，帮助建立了植物生理学。一些机械论者甚至希望改进的显微镜能使人直接观察到原子及其形状和运动，从而展示机械论哲学的基本解释原则。

和所有其他观察一样，显微镜观察也会遇到相互冲突的解释。可以把精子的发现解释为对预成论的支持。当时的人发现，腐水中会自然产生大量生物，这强烈支持了已有的自然生成说，即生物可以从非生命物质中出现；而这又支持了一种渐成论观念，即生命结构可以从原本无组织的物质中产生。数个世纪以来，自然哲学家大都认为简单生命是在特定环境下自然出现的，比如腐烂的牛的尸体产生蜜蜂，泥浆产生蠕虫，腐肉产生蛆虫。在17世纪60年代于美第奇宫廷完成的一系列著名实验中，弗朗切斯科·雷迪（1626—1697）将几块肉放至腐烂，有的覆上网孔或布料，有的露天搁置。露天搁置的肉生了蛆，而防止苍蝇接近的肉则没有生蛆。和大多数事后认为的"决定性"实验一样，雷迪的实验并未立即消除自然生成说，因为该事实还可以作（也的确作了）其他解释。雷迪本人也承认，某些昆虫——如橡树瘿蜂——或许是从植物中直接产生的。现代人常常嘲笑自然生成说，但需要指出的是，当时任何近代科学家只要不相信第一个生命形式是通过上帝的奇迹介入而特创的，就只能相信生命是从非生命物质中自然生成的。

不论是显微镜还是对生命系统的机械论看法均未实现预期结果。受当时可资利用的材料和光学系统的制约，显微镜很快就达到了放大率和分辨率的极限。显微镜研究表明生命系统极为复杂，机械论解释越来越难以解释生命的形成或功能。然而即使在机械论最盛行之时，也出现了许多更具生机论色彩的模型。事实上在17世纪，生命与非生命的区分并不清晰，许多思想家都将机械论体系与生机论体系混合在了一起。例如，很少有机械论者会严格到否认生命系统中存在着赋予活力的灵魂。这种灵魂并不必然类似于基督教神学所说的那个非物质的、不朽的人的灵魂，而是被认为以各种形式或层次存在于各种实体中［对现代读者而言，也许用"生命精气"（vital spirit）一词能够更好地表达这个概念］。这些观念可以追溯到亚里士多德，他曾经提出灵魂的三种层次：植物中的**营养**灵魂，负责生长和营养吸收；动物除了营养灵魂还有一种**感觉**灵魂，负责感觉和运动；人除了营养灵魂和感觉灵魂之外还有一种**理性**灵魂，负责思维和理性。在许多人看来，虽然机械原则可以解释特定的身体功能和结构，但是整个有机体——更不要说意识和知觉——需要灵魂来组织和维持。

海尔蒙特的学说

17世纪出现的最全面的新医学体系也许是佛兰德斯的贵族、医生、化学家和自然哲学家海尔蒙特（1579—1644）的学说。海尔蒙特将化学、医学、神学、实验和实践经验结合成一个连贯的、极有影响的体系。他的自述表达了对传统学问的不满和对追求新知的渴望，这是科学革命时期思想家的典型态度。他讲述了自己如何进入鲁汶大学，然后又因为觉得学无所成而没有接受学位。他又向一些耶稣会士学习，还是一无所获。接着他获得了医学博士学位，却发现医学的基础已经"腐烂"，而后又转向帕拉塞尔苏斯的学说，最终还是拒绝了它的大部分内容。海尔蒙特因此决心从头开始，自称"火的哲学家"，意指其训练并非来自传统学问，而是来自化学熔炉中的实验。的确，海尔蒙特是一位杰出的观察者，他描述了多种疾病的起源、症状和病情发展，如果没有他，这些疾病要几个世纪后才能得到认识。

海尔蒙特拒绝接受亚里士多德的四元素理论和帕拉塞尔苏斯所说的三要素，他宣称水是构成万物的唯一元素。这个想法不仅让人想起已知最早的希腊哲学家泰勒斯，而且（在海尔蒙特看来）更重要的是，它会让人想起《创世记》1: 2所说的，神的灵"[像母鸡一样]覆在**水**面上"而产生了世界。这位比利时哲学家试图用实验来确证这种想法，最著名的便是柳树实验。海尔蒙特在200磅的土壤中种了一棵5磅重的柳树苗，为它浇水5年，最后柳树重164磅而土壤重量基本没有减轻。海尔蒙特由此得出结论，认为树的整个构成必定只来自于水。根据海尔蒙特的说法，创世时植入世界的种子（*semina*）能把水转变成任何物质。这些种子并不是像豆子那样的有形种子，而是非物质的组织原则，如同把蛋黄液组织成一只鸡的那种无形的生命本原。火和腐烂过程摧毁了种子及其组织能力，把原来的物质变成了类似空气的物质，海尔蒙特称之为“气体”[*Gas*，这个词来自于chaos（混乱），我们用来表示第三种物态的词便是直接来源于它]。于是，燃烧的木炭和发酵的啤酒会散发出令人窒息的林木气，燃烧的硫磺会散发出带有恶臭的硫磺气。在大气层的寒冷部分，这种气体复归为原始的水并下落成雨，从而完成了水在海尔蒙特的自然体系中的相继变化所构成的循环。

海尔蒙特主张身体过程本质上是化学的，这一观点类似于帕拉塞尔苏斯但更为精致。他认识到胃液的酸性是消化的原因，并且对体液作了分析，特别是通过分析尿液来寻找肾结石和膀胱结石的病因和治疗方案——结石曾是17世纪最令人痛苦的疾病之一。然而，单单是化学过程并不足以解释生命过程，它们还必须依靠寓于体内的一种类似于精气的东西即阿契厄斯（*archeus*）的引导。对海尔蒙特而言，阿契厄斯调节和管理着人体的机能。疾病缘于虚弱的阿契厄斯无法履行自身的职责，因此医学上的治疗必须着眼于增强阿契厄斯的力量。于是，海尔蒙特拒绝接受盖伦所说的体质观念、四体液和治疗方法。他认为，像瘟疫这样的疾病并非由于体液失衡，而是由于外界的疾病“种子”侵入和改变了身体。强大的阿契厄斯可以驱散这些种子，而虚弱的阿契厄斯则需要帮助。(注意，在盖伦和海尔蒙特的医学中，医生的作用都是**辅助**自然过程，而不是使自然过程转向或者对身体加以控制。)海尔蒙特也强调病人心理状态和情绪状态的作用，并且声称，想象力可以引起身体的生理变化。海尔蒙特的观念深刻地影响了许多医生、生理学家和化学家。

机械论和生机论的生命观并非不可调和，而是处于一个连续谱系的两端，许多医生和自然哲学家都采取了中间立场。与海尔蒙特同时代的伽桑狄也认为，种子是能够重新组织物质的强有力的本原。但海尔蒙特的种子是非物质的，而伽桑狄的种子则是机械地作用于物质的（由上帝组织的）物理原子的特殊结合。事实上，机械论和生机论的思辨在18世纪产生了混合的医学体系，例如格奥尔格·恩斯特·施塔尔（1659—1734）的体系既强调化学转变的机械性质，又要求生命力对生命系统进行组织和控制。赫尔曼·布尔哈夫（1668—1738）也许是18世纪医学尤其是教学法方面最有影响的人物，他将17世纪自然哲学的不同潮流融合在一起。作为莱顿大学医学院的化学和医学教授，布尔哈夫既大力倡导希波克拉底的治疗方法（强调环境和病人的个人特征），也强调化学对于医学教育的重要性。他研究医学和人体的进路结合了玻意耳的机械论哲学、牛顿物理学和海尔蒙特的“种子”说。布尔哈夫的医学教育改革被欧洲许多地区所采用（因此他有时被称为“欧洲之师”），并为18世纪医学的重要转变奠定了基础。

植物和动物

对植物和动物的研究——我们今天所谓的植物学和动物学——在16、17世纪大大扩展。这类材料的传统文本出自一种百科全书传统，它源自老普林尼（23—79）为了搜集希腊学术并向罗马公众普及而编写的巨著《博物志》。关于动植物的百科全书式论述充满中世纪的本草志和动物寓言集，这类著作一直持续到科学革命时期。其中最著名的之一是康拉德·格斯纳（1516—1565）所著的五卷本《动物志》，其中附有数百幅木刻画。然而，现代读者会觉得这类著作很怪异，因为它们将关于各物种的自然主义细节描述，与自古以来针对每一种动植物累积起来的大量文学、词源学、圣经、道德、神话学和隐喻含义混杂在一起。倘若描述孔雀不提它的傲慢，说到蛇不提它在亚当堕落中扮演的欺骗者角色，提及车前草（一种生长在人行道旁的普通植物）而未指出它意味着基督常走的道路，那么这样的描述必定是不完整的。它并未把动植物呈现为孤立的物种，而是将其置于一个丰富的意义和典故网络之中。动植物既是自然物又是象征，后者依赖于对世界的想象，这个世界有多层意义，**既是**由字面意义**又是**由隐喻意义构成的，充满了有待解读的象征性寓意。因此，得到描述的不仅有人所熟知的生物，也有传说中的动物，如独角兽、龙和各种怪兽。这并不必然是因为作者相信它们出没于地球，而更多是因为它们存在于文学世界因此承载着意义，无论它们是否存在于自然界。现代读者也许会认为这些文本“离奇”、轻信或充斥着“非科学的琐事”，但当初的读者很可能会认为现代植物学和动物学的描述性文本枯燥乏味，而且古怪地脱离了与人类的关系。

近代早期有两项进展使这一象征传统转移到了别的方向。首先是医学需要辨认草药。随着人文主义学者继续复原、编辑和出版希腊医学文献，人们越来越需要识别这些文献中提到的药用植物，并且确定它们在野外的生长方位。因此就需要新的本草志将古代文献与16世纪田野中生长的植物关联起来。为此，新的本草志不仅将草药的常用名与其古希腊名称相联系，而且为其绘制了准确的自然主义插图。就像维萨留斯与提香工作室中的艺术家们合作一样，16世纪的新一代植物学家和艺术家合作完成了配有大量写生插图的本草志，例如奥托·布伦费尔斯的《本草活图》（1530—1536）、莱昂哈特·富克斯的《植物志》

（1542）。另一项进展是欧洲人视野的拓宽。在最狭窄的层面上，迪奥斯科里季斯等古代权威描述的主要是地中海地区的植物，而并未认识北欧的物种，因此有必要对没有古典词源系谱的植物作出描述。在欧洲以外（尤其是美洲）航海首次遇到的无数动植物也有同样的问题，但范围要大得多。马铃薯、玉米、西红柿等食用植物，"金鸡纳树皮"（奎宁的来源，可治疗疟疾）等药用植物，以及负鼠、美洲虎、犰狳等新发现的动物，大大扩展了欧洲人知晓的动植物种类。这些新物种没有建立起对应性和象征性的网络，无法纳入传统的本草志和动物寓言集。来自新世界的大部分报告首先到达西班牙，那些应国王之嘱组织信息的学者不得不放弃了基于普林尼等古典模式的传统百科全书方法，不仅因为新发现使传统分类过时了，还因为持续涌入的新的信息使学者们不可能对这些知识进行全面的整理。

新世界的西班牙人往往是修会成员，他们试图记录当地的植物、动物和医疗活动，有时会与当地学者合作编写插图文本。有时被称为"新世界的普林尼"的何塞·德·阿科斯塔（1539—1600）是秘鲁的耶稣会士，他不仅创建了五所学院，还写了一本拉丁美洲的博物志，该书被译成多种语言，在欧洲广为流传，备受引用。1570年，西班牙国王菲利普二世派他的医生弗朗西斯科·埃尔南德斯随探险队前往新世界专门寻找药用植物。埃尔南德斯花了七年时间（主要是在墨西哥）为植物编目，并且向当地治疗师询问它们的药效，一批当地艺术家则为六卷本的《新西班牙的动物和植物》绘制了大量插图（该书描述了大约3000种植物和数十种动物）。由于实在无法将新的植物纳入古典分类方案，埃尔南德斯甚至采用当地名称来创建一种新的植物分类学。与此同时，方济各会修士贝尔纳迪诺·德·萨阿贡（1499—

1590）在数位阿兹特克助手和信息员的协助下在墨西哥特拉特洛尔科的圣克鲁斯学院完成了《新西班牙风物通志》。这是一部用西班牙语和纳瓦语双语写成的长篇巨著，描述了阿兹特克人的文化、习俗、社会和语言。在西班牙本土，医生莫纳德斯（1493—1588）编写了《西印度风物医药志》，描述了数十种来自新世界的物种。加西亚·德·奥塔（1501—1568）和克里斯托旺·达·科斯塔（1515—1594）等葡萄牙学者也描述了他们在印度以及东亚和南亚其他地区新发现的动物和药用植物。

对新药物的寻求促进了对新植物的研究，也因此促进了植物园的建立，它们通常附属于大学的医学院。在整个中世纪，药用植物园一直是修道院的一部分，新的植物园在此基础上建立起来，并且为了教学和研究的目的而扩展。第一批植物园于16世纪40年代出现在意大利的比萨大学和帕多瓦大学，1568年出现在博洛尼亚大学，随之建立的还有药用植物学的教席。其他医学教育中心相继建立了自己的植物园，例如巴伦西亚大学（1567）、莱顿大学（1577）、莱比锡大学（1579）、巴黎大学（1597）、蒙彼利埃大学（1598）、牛津大学（1621）等等。这些植物园以严格的秩序进行安排，依照药性、形态学或地理起源将植物进行分组。植物的种子、根、插条和球茎被人们寻求、购买和交换，全欧洲植物园中的植物种类因此得以扩充。私人也开始对珍稀植物的栽培和育种感兴趣，从而引发了17世纪荷兰著名的“郁金香狂热”。在荷兰，新兴的中产阶级投入巨资购买珍稀品种，艺术家则通过静物写生来保存这些奇花异草。

对异国珍稀品种的广泛兴趣也反映在“珍奇馆”（图15）所收藏的所有种类的自然标本中。这些收藏在某种意义上是博物馆的前身，能够显示收藏者的权力、财富、人脉和兴趣，也能激

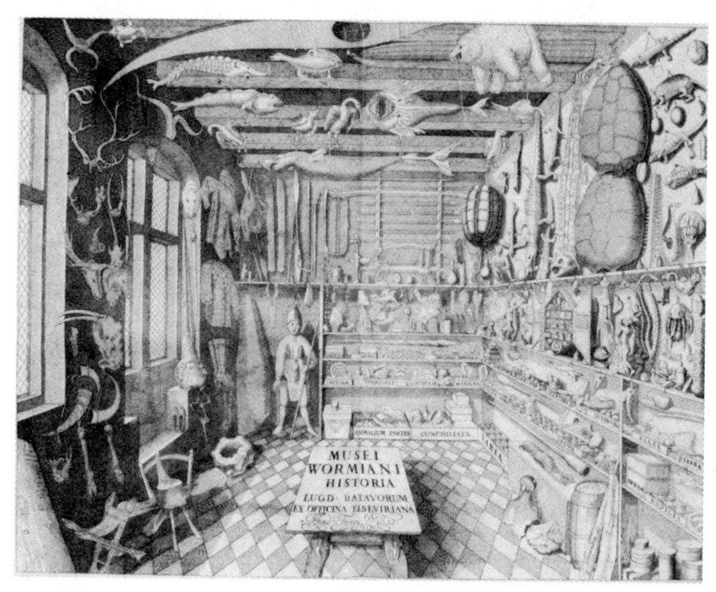

图15 奥勒·沃姆的珍奇馆。出自《沃姆博物馆》(莱顿, 1655)。

起人们对自然和人工奇迹的惊叹。王公贵族和学者们积累的收藏既包括自然物，如动物、植物和矿物标本，也包括人工物，如机械发明、绝妙的手工艺制品以及人种学和考古学物品。乌利塞·阿尔德罗万迪(1522—1605)是最早收集此类藏品的人之一(部分藏品仍然藏于博洛尼亚大学)，基歇尔在罗马学院的博物馆(他亲自做导游)是17世纪罗马游客的“必访之地”。展柜中物品的排列侧重于物品之间的关联，这些关联常常出乎我们的意料。于是，这些展柜成了另一种小宇宙，一室之内便可展示和象征相互关联的人与自然的多样性、奇异性和异国风情。

第六章

科学世界的建立

科学不只是关于自然界的研究和知识积累。从中世纪晚期至今，科学知识被越来越多地用于改变这个世界，赋予人类更大的能力控制世界，并且创造出新的世界让我们居住，我们似乎与自然界渐趋疏离。现如今，人类日益被技术创造的人工世界所包围。只有当技术出问题时，我们才发现自己是多么依赖技术，此时我们就像中世纪的农民面对久旱无雨一样感到无助。因此，当自然界通过侵扰这个人工世界来重新彰显自己的力量，比如陨石或太阳耀斑干扰卫星通信、雷击切断电能，或者火山爆发使飞机停运时，现代人往往会惊恐万状。在过去几个世纪里，技术扩张最彻底地改变了人类的日常世界。与此同时，技术的发展也依赖并且促进科学上的探索。16、17世纪发生了一次特殊的转向，即用科学研究和科学知识来解决当时的问题，满足当时的需求。

人工世界

在文艺复兴时期的意大利，新的宏伟工程改变了自然和城市的面貌。运河和供水系统占用了新的土地，提供了饮用水和运输路线。菲利波·布鲁内莱斯基（1377—1446）用新的建筑技术为大教堂建造了一座巨大的双层圆顶，重塑了佛罗伦萨的天际轮廓。新的城市设计体现了人文主义者对公民生活、对执政君主智慧与权力的强调，新的防御工事则用来保护他们的利益。通常情况下，一项新技术会推动其他技术的发展。军事技术在15世纪发生了转变，火药的使用日益增多，轻便的青铜大炮也被制造出来，这一切都使中世纪的防御工事变得过时——其屹立的城垛正是火炮的极好目标。因此，必须发展一种新的防御系统。新的防御设计利用了几何原理，成为贵族教育的必修内容。先是在16世纪的意大利，然后在其他地方，迫切的实际问题（以及君主的野心）造就了一个由博学的工程师和建筑师所组成的群体，他们以古代的阿基米德和维特鲁威为榜样，愈发倾向于运用数学原理和分析来解决实际问题。这一新兴群体既非只重手工经验积累的工匠，亦非远离实际事务的学者，而是介于他们之间。科学革命的一个基本特征是越来越多地利用数学来研究世界，恰恰是这批人为此提供了关键的背景。达·芬奇（1452—1519）和16世纪中叶的军事工程师塔尔塔利亚都是这一"中间"群体的早期例子。到了16世纪末，伽利略从这些博学的工程师那里汲取了灵感，借鉴了方法。

改建罗马城的灵感来自于实用性和效仿古人的人文主义渴望。教皇资助研究和重建了古老的水道和排水系统。建于公元4世纪的旧圣彼得大教堂被推倒重建，宏伟的新圣彼得大教堂矗立至今。此举激励了16世纪的一项宏伟的工程——移动梵蒂冈的方尖碑。这是罗马人在公元1世纪竖立的一块六层楼高的巨石，重逾360吨。1585年，由于方尖碑妨碍了新的圣彼得大教堂的建设，教皇西克斯图斯五世悬赏征集方案，将这块古埃及巨石移到新的位置。这是方尖碑在1500年里第一次移位。最后，工程师多梅尼科·丰塔纳（1543—1607）竞标成功。1586年4月30日，利用75匹马、900个人、40台卷扬机、5个50英尺长的杠杆和

图16　移动梵蒂冈的方尖碑，出自多梅尼科·丰塔纳同名作品（罗马，1590）。

8英里的绳索，丰塔纳成功地将这块装在铁架内的巨石抬离了基座。教皇对这项工程异常重视，甚至不惜将新落成的一部分圣彼得大教堂拆除，以使杠杆和卷扬机最好地发挥作用。接着，丰塔纳将方尖碑放低置于一个移动托架之上（图16），沿堤道运走，将其重新竖立于今天的所在地——圣彼得广场中心。

文艺复兴时期的成就及其背后的经济和军事动力都需要原料。因此，1460至1550年见证了采矿业的繁荣，特别是在矿产资源最为丰富的中欧地区。中世纪的采矿主要是开采地表矿藏等小规模活动。但近代早期欧洲的需求——武器和火炮需要铁和铜，造币需要白银和黄金——催生了更有条理、更大规模的采矿以及更好的冶炼、精炼技术。更深的竖井和更大的规模除了要求组织更多的劳力外，还要求更多的机械化——用水车驱动风箱和破岩设备，用泵为矿井排水和通风。德国人文主义者和教育家乔治·阿格里科拉（1494—1555）也许是最有名的记述采矿的作者，他希望对采矿知识进行整理和改进。为使这项原本肮脏的行业受到敬重，他撰写了富含插图的拉丁文巨著《论金属》，把德国采矿活动与古典文献联系起来，并且为冶金学创造了一套拉丁文词汇。阿格里科拉的插图中偶尔出现的伐木、烟尘和废水横流等场景，预言了此种技术持续发展将会伴随着何等惨痛的环境代价。对实际从业者更有用的也许是采矿业的监督人埃尔克（约1530—1594）的德语著作。他的著作论述了如何实际处理矿石、测定金属，制备酸类和盐类化学产品，包括火药的最重要成分硝石。采矿业到16世纪中叶便衰落了，既是因为欧洲矿藏的枯竭，也是因为来自新世界的金属压低了金属价格，使欧洲采矿业已不那么有利可图。

对新世界的期待促进了制图学和航海的发展。中世纪晚期的航海图，或称波尔托兰海图，只标明了海岸线，一些特定的点覆盖着玫瑰花形的罗盘航向。这些图可用于地中海地区或沿海岸线的相对较短的航行，但不具备地理投影效果，也不能用于越洋航行。托勒密写于公元2世纪的《地理学》在15世纪被重新发现，它用东西线和南北线（分别为经度和纬度）所组成的网格来绘制地图。15世纪末的地图——如瓦尔德泽米勒的地图——采用了这种方法，图中弯曲的纬度线和经度线向两极会聚。佛兰德斯制图师墨卡托（1512—1594）普及了如今广为人知的墨卡托投影法，其中经线相互平行，并与垂直的纬线相交成直角。虽然它扭曲了高纬度地区的大陆块，但这种把球形地球投影到平面地图上的方法更便于导航（至少在低纬度地区），因此深受西班牙宇宙志学者和航海家的青睐。

罗盘和象限仪——分别用来确定航向和纬度——自中世纪以来就被用于航海，但确定经度的可靠方法尚付阙如。当船在欧洲水域或陆地视线范围内航行时，这尚不构成严重的问题。但如果没有准确的经度测量，越洋航行将非常危险。由于定位同时需要纬度和经度，对于制图员和航海家来说，缺少经度是十分严重的问题，找到测定经度的方法已经成为这一时期最迫切的技术问题。西班牙、荷兰、法国和英格兰等相互竞争的航海国家都高额悬赏，征集测定经度的可靠方法。

测时是测定经度的关键。两个地方的本地时间每相差一小时，经度就相差15度（因此一个现代“时区”大约宽15度）。然而，如何同时知道两个遥远地点的时间呢？可以带着在船的始发地设定的时钟，将其读数与由观测太阳或星星的位置所确定的船所在地的时间相比较。不幸的是，近代早期时钟的误差有每天20分钟之多。伽利略观察到，不论振幅多大，摆的节奏是恒定的，这暗示了一种新的计时调节器。他在软禁期间开始设计一种由摆调节的时钟，但从未制造出来。1656年，荷兰的克里斯蒂安·惠更斯制造出了第一个可使用的摆钟，使可靠性大幅提升，至少对于陆地上的时钟是如此。在颠簸的船上，摆钟无法精确运行。此后，惠更斯和胡克各自独立试验了以弹簧为动力的时钟，但事实证明，它们在船上走得也不够精确。不过，胡克在研究弹簧的过程中提出了弹簧的拉伸与受力之间的关系，即今天所谓的“胡克定律”，惠更斯的工作也使简谐运动定律得到了改进。（经度问题本身直到18世纪才得到解决：英格兰仪器制造者约翰·哈里森设计出一种新的精密计时器，使时钟即使在海上也能精确运行。）

除了人造时钟还有天体钟，即某个天文事件，它在参照位置的发生时间可以计算出来，然后与该事件在观察者所在位置发生时的本地时间相比较。16世纪的西班牙宇宙志学者通过对月食的坐标观测，成功地测定了西班牙帝国殖民地的经度，但月食对于航海来说太过罕见。然而，木星四颗卫星的食发生得更为频繁，最里面的卫星木卫一每42小时就发生一次食，伽利略建议把它们用作计时器。天文学家吉安·多梅尼科·卡西尼（1625—1712）最充分地探索了这种想法，于17世纪60年代编制了这些食的时间表。然而，虽然这个系统在陆地上管用（它曾被成功地用于修正陆地地图），但事实再次证明，在移动的船上用望远镜观测食是不现实的。不过在检验这种想法的过程中，观测者们注意到，一些食的发生时间要比预计的晚几分钟。丹麦自然哲学家奥勒·罗默（1644—1710）注意到，当木星距离地球最远时，这一误差达到最大，遂于1676年提出光速是有限的（食的表观延迟缘于光在空间中的传播时间），从而使粗略测量光速成为可能。

这寥寥几个例子表明，技术应用与科学发现有着千丝万缕的联系，两者相互驱动和促进。“纯粹”科学与“应用”科学的对立并不适用于17世纪。如果贬低实际需要——无论是军事、经济、工业、医疗还是社会政治方面的需要——对于促进科学革命发展的重要性，将会与真实的历史情形背道而驰。

一提到科学发现与实际应用的关联，人们往往会想到弗朗西斯·培根爵士（1561—1626）。培根出生于一个社会地位很高的家庭，受过律师教育，曾任国会议员，被授予维鲁拉姆勋爵爵位，最终担任了英国上议院的大法官（因涉嫌受贿被罢免），其一生中的大多数时间都居住在权力的宫殿之中。因此，关于权力的话题和帝国的建筑一直在他的思考范围中。他主张自然哲学知识应当**为人所用**，因为它能够为人类和国家增添福祉。培根将当时的自然哲学刻画为——或讽刺为——毫无价值，它的方法和目标被误导，从事自然哲学的人终日陷入语词之争，对于具体工作却视而不见。事实上，虽然培根对自然魔法的形而上学基础表示了怀疑，但他称赞了魔法，因为魔法“主张要把自然哲学从各种思辨召回到重要的具体工作”。自然哲学应当具备**操作性**而非思辨性——应当做事情，制造东西，赋予人类以力量。他认为，印刷术、指南针和火药等所有技术成果构成了人类历史中最具变革性的力量。因此，培根呼吁“彻底重建科学、艺术和所有人类知识”。

方法论对于培根渴望的变革至关重要。他主张编纂“博物志”，即广泛收集观察到的现象，无论这些现象是自发出现的还是人为实验的结果，即他所谓的强迫自然偏离其日常进程。充分收集原始材料之后，自然哲学家就可以将其结合在一起，通过归纳过程提出越来越普遍的原理。关键是要避免过早提出理论、进行纸上谈兵的思辨和建立宏大的解释体系。一旦发现更一般的自然原理，就应当富有成效地运用它们。然而，培根并非主张一种完全的功利主义。实验不仅在产生结果（实际应用）时有用，在启迪心智时也有用。真正的自然知识同时致力于“荣耀造物主和抚慰人的处境”。虽然培根明确指出，壮大和扩张英国是其事业的一个目标——他请求给予他的改革思路以国家支持，但伊丽莎白一世和詹姆斯一世都没有作出回应——但在更长远的意义上，培根认为这种操作性知识旨在重新获得神在《创世记》中赐予人类、却随着亚当的堕落而失去的统治自然的权力。

至关重要的是，培根思索的不仅是自然哲学的方法和目标，还有其体制结构和社会结构。他坚持认为，必须用合作的集体活动取代独自做学术研究的过时理想。事实上，培根所制定的事实收集方案需要大量劳动力，虽然他本人也着手进行收集，但他所能完成的非常少。培根晚年在一则乌托邦式的寓言《新大西岛》（1626）中表达了他所设想的改革后的自然哲学以及由此可能创建的更好的社会。这个故事描述了太平洋上一个和平、宽容、自给自足的基督教王国——本色列岛。这个岛上的生活之所以幸福，不仅是因为有贤明的君主，更是因为有所罗门宫，这是一个由国家资助的自然研究机构，致力于“认识事物的原因和秘密的运动；拓展人类王国的疆界，实现一切可能之事”。所罗门宫的成员集体研究自然，尽管有劳动分工和等级安排——较低等级的收集材料，中间等级的做实验和进行指导，最高等级的进行解释。在本色列岛，培根式的自然哲学家形成了一个受政府支持的、备受尊敬和有特权的社会阶层，服务于国家和社会。当17世纪欧洲的许多自然哲学家正在为其社会地位而努力抗争时，培根的愿景无疑令他们感到鼓舞。

科学社团的兴起

今天，科学研究无所不在，其中一些研究与所罗门宫的某些特征甚至不无相似之处。科学家的工作场地有大学，有政府的、企业的和独立的实验室，有大型特殊仪器（如望远镜或粒子加速器）的所在地，还有野外、研究站、动物园、博物馆等等。单个的科学家经由专业组织、科学团体和科学院、研究小组、通信以及新兴的互联网结合成一个社会群体。科研经费来自政府研究资助、企业的研究和开发部门、大学以及私人捐助。物理场地、社会空间和赞助这三个特征对于现代科学的运作必不可少。这些特征在科学革命时期的确立对于构建我们今天所知的科学世界至关重要。从整个17世纪一直到18世纪，自然哲学家的工作变得越来越正式化。个人结合成私人协会，私人协会又逐渐演变成国家科学院。个人的通信交流发展成为出版的期刊。自费的业余研究者和以大学为基础的自然哲学家结合成第一批拿薪水的专业人士。

在中世纪晚期，自然哲学研究主要在大学、修道院以及——较小程度上——在少数宫廷中进行。这些传统活动场地在16、17世纪仍然重要，但有了新的补充。对于文艺复兴时期的人文主义运动来说，在大学之外建立学术群体至关重要。在这些群体中，学者与志同道合的人分享他们的工作，获得支持、批评和认可以及偶尔的资助。这些早期群体大多是文学或哲学性的。而到了16世纪末，自然哲学家将模式扩大，由此兴起了第一批科学社团。最早的科学社团创立于意大利，17世纪成立了数十个，数量多于欧洲所有其他地方。不过其中大多数社团都是地方性的，留存时间也不长。

其中最早的社团之一是猞猁学院。它的名字暗指目光敏锐、擅于洞察的猞猁。该学院由切西亲王（当时是一个18岁的罗马贵族）和三位同伴1603年创立于罗马，维持了大约30年。切西创立猞猁学院时坚信，研究自然是复杂而费力的事情，需要集体努力。猞猁学院的成员并不多，但其中包含了德拉·波塔、伽利略、斯滕森和约翰·施莱克等拥护自然魔法的人，施莱克后来成为耶稣会传教士，把欧洲科学知识带到了中国。猞猁学院成员致力于研究自然哲学的所有分支，这些研究往往是独立进行的，但偶尔也有合作，比如他们根据已从西班牙带到意大利的埃尔南德斯考察墨西哥的手稿，长期致力于出版《来自新西班牙的药物》（1651）。猞猁学院成员倡导用新的化学方法来研究医学，支持伽利略的工作（他1613年的《关于太阳黑子的书信》和1623年的《试金者》都是在猞猁学院的资助下出版的），还做了显微镜研究。猞猁学院因切西在1630年的过早离世而失去了领袖和靠山，不久便解散了。

1657年，在很大程度上是由于莱奥波尔多·德·美第奇亲王对自然哲学的个人兴趣，西芒托学院在佛罗伦萨的美第奇宫廷成立了。它的座右铭“通过检验和再检验”（*Provando e reprovando*）概括了该群体对实验的专注。美第奇宫廷提供了一个集体研究中心，这是猞猁学院所缺乏的，美第奇的资助使它能够持续运行。许多成员都是伽利略的追随者，他们继续了伽利略的许多研究计划和方法。不过，这个佛罗伦萨学院的成员对从解剖学和生命科学到数学和天文学的各种东西都进行研究，而且特别重视研究和改进新仪器，如气压计和温度计，莱奥波尔多本人也参与其中。雷迪、马尔比基、博雷利等许多著名意大利自然哲学家的工作都是在西芒托学院完成的。由于成员之间的分歧，几位杰出人物的离去，加之莱奥波尔多被提名为红衣主教，从而不得不花更多的时间在罗马，这一切导致西芒托学院于1667年关闭。存在仅10年的西芒托学院是自然哲学家自愿联合起来对自然进行集体实验性研究的最引人注目的典范。

到了17世纪中叶，科学社团传播到了阿尔卑斯山以北。1652年，四名德国医生成立了自然奥秘学院。早年间，这个“自然探究者的学院”主要专注于医学和化学主题。该学院1662年发布的章程声明其目标是“荣耀上帝，启蒙医术，惠及同胞”。学院发展十分迅速，虽然其成员遍及德语区各处，无法作为一个团体定期会面，但尤其是通过（从1672年开始）每年出版一卷由各成员提交的论文，自然奥秘学院致力于将他们实际联系在一起。1677年，神圣罗马帝国皇帝利奥波德一世给予它正式认可。在随后若干年中，学院17世纪的建制不断扩张，远远超出了医学和生命科学，并最终发展为今天的德国科学院。

17世纪50年代，一个被称为“实验哲学俱乐部”的群体开始在牛津大学的沃德姆学院会面讨论自然哲学，用机械装置进行实验，观察解剖和演示。克里斯托弗·雷恩和胡克都是其早期成员，后来玻意耳等17世纪中叶的其他英国知名人士也加入进来。查理二世1660年王政复辟后，该俱乐部的几位成员和其他一些人为一个更加正式的法人组织制定了章程，并于1662年获得皇家特许状，成为改进自然认识的伦敦皇家学会。皇家学会一直持续至今，标志着科学社团发展的一个新阶段。和西芒托学院一样（皇家学会一直与它保持着通信），共同做实验也是皇家学会的重点，但皇家学会被视为一个大得多的、更为正式的组织。200多名会员很快被选出，其中大多数人是英国贵族，这种选择反映了对他们提供财政支持（而不是思想贡献）的一厢情愿的期待。皇家学会明确以培根及其指示为榜样，自行设定公众目标和社会目标。事实上，可以把皇家学会看成实现所罗门宫的尝试。许多早期会员都曾参与内战期间的乌托邦计划、教育方案和企业规划，并把这些目标带到了皇家学会。他们严格避免教派和政治上的依附，希望能在自然哲学中找到一种共识作为基础，克服之前内战期间的派系纷争。

皇家学会会员在伦敦的格雷欣学院定期举行会议，在那里做实验，展示新的研究和观测成果。当时（以及此后）几乎所有著名的英国自然哲学家都是皇家学会会员。皇家学会的成员很快就超越了英国国界，无论当时还是现在，当选为会员都会带来极大声望。也许早年最重要的创新是皇家学会秘书亨利·奥尔登堡1665年创办了第一份科学期刊——《哲学会刊》。创办《哲学会刊》最初始于奥尔登堡的私人努力，他徒劳地希望靠预订收入维生，但《哲学会刊》很快就变得在概念上与英国皇家学会关联起来，虽然直到后来两者才有了正式关联。奥尔登堡维持着大量通信（因此他曾被当作间谍囚禁在伦敦塔中），从而能够报告整个欧洲的科学进展。《哲学会刊》不仅发布英国皇家学会的活动，而且发表国外的报告、科学书信和书评。尽管以英语为主，但它成为欧洲科学生活的一个重要发声途径——学者们可以在这里发表评论，公布调查结果，确立优先权和进行争论。牛顿关于光、光学和他的新望远镜的论文都发表在这里，列文虎克的显微镜发现是从荷兰邮寄来的，马尔比基的解剖研究则发自意大利。关于彗星的争论以报告的形式在这里争相发表。每当玻意耳有某种东西要简要报告，就会发行专号。

尽管雄心勃勃，但英国皇家学会也遭遇到早期科学社团常见的一些问题——重要成员的流失、资金短缺、缺少赞助，等等。它的许多宏伟计划到头来都化为了泡影。大多数成员都不够积极，只是偶尔交纳会费或根本不交，王国政府给予学会的唯一礼物就是“皇家”这个形容词。其改进贸易的培根主义计划举步维艰，因为商人们不愿分享他们的专有知识和技能——这是情理之中的事。自然哲学界以外的英国人的回应也好不到哪儿去——托马斯·沙德韦尔的喜剧《大师》（1676）嘲讽了皇家学会、学会会员及其活动，乔纳森·斯威夫特的讽刺小说《格列佛游记》（1726）中的“勒皮他岛游记”尖刻地模仿了他们关于公用事业的主张。奥尔登堡在1677年的去世导致《哲学会刊》一度停办，玻意耳在1691年的去世意味着皇家学会失去了其最为活跃和慷慨的会员。牛顿从1672年开始担任会员，1703年成为皇家学会会长，那时他被视为英国卓越的自然哲学家。他的威望为皇家学会注入了新的活力，但他往往喜欢那些对他本人的研究有促进作用的工作，这使皇家学会此前活动的广度有所缩减。不过到了18世纪中叶，皇家学会的地位变得越来越稳固，并且一直持续至今。

不同于英国皇家学会自下而上的建立，巴黎皇家科学院是自上而下建立的。它源于路易十四的财政部长让—巴普蒂斯特·科尔贝（1618—1683）的设想。科尔贝既是为了给支持艺术与科学的太阳王增添荣耀，也是为了以有益于国家的方式将科学活动集中起来——这是路易十四长期统治期间规模更大的中央集权化的一部分。皇家科学院于1666年举行了第一次会议，有20位院士参加，从荷兰招募的惠更斯任院长。他们在国王图书馆每周会面两次，以期进行合作研究（这并不总能顺利进行），并获得薪水和研究资助。由此，法国人比英国人更好地实现了培根的设想。作为对王室拨款的报答，院士们应当为国家问题找到科学的解决方案。两位薪水最高的成员——惠更斯和卡西尼——都在研究经度这一重要问题时被带到法国，这绝非巧合。院士们还检测了凡尔赛宫和整个法国的水质，评估了新的项目和发明，考察了书籍和专利，在皇家印刷厂等处解决了技术问题，并且第一次对法国进行了准确的勘测。最后的勘测活动表明，法国面积比此前认为的要小一些，据说路易十四对此说了一句妙语，即就减少王国领土这一点而言，他自己的院士成功了，而他所有的敌人都失败了。然而，尽管服务于国家，院士们还是有足够的时间从事其他研究，特别是他们为自己制定的几个大的集体项目，包括编写详尽的动植物博物志（图17）。

皇室赞助还为院士们提供了工作区：一个化学实验室、一个植物园和（当时）位于巴黎郊区的一个天文台。1672年落成的巴黎天文台起初打算作为整个科学院的一个住所，后来成了天文学家的专属地盘。在高额的薪水和掌管新天文台的诱惑下，天文学家卡西尼辞去了为教皇服务的职位来到巴黎，天文台完工之前就住到了那里。卡西尼和他之后的三代人使巴黎天文台成为欧洲最重要的天文学机构。它的南北中心线标志着地球的本初子午线，从那里进行的经度测量已有两个世纪，直到1884年本初子午线被确定为通过格林威治的线。（格林威治的皇家天文台成立于1675年，特别旨在"确定经度以改进航海和天文学"，此时巴黎天文台刚刚成立不久。）皇家资助也使巴黎科学院得以向国外派出科学远征队——到圭亚那、新斯科舍和丹麦进行天文观测，到希腊和黎凡特收集植物标本，尤为著名的是18世纪初到南美洲和拉普兰进行观察和测量，以检验笛卡尔和牛顿关于地球精确形状的预言。它还收集和发表了暹罗、中国等地的耶

图17 巴黎皇家科学院的院士们在进行解剖。秘书（让—巴普蒂斯特·迪阿梅尔）记录着观察结果，几位院士在讨论；窗外可见国王花园。《作为自然历史的动物》（海牙，1731；初版于巴黎，1671）。

稣会士发来的评论，并与英国皇家学会会员（即使在英法战争期间）和整个欧洲的其他学者进行了广泛的通信。

科学院以外的科学群体

1700年以后，科学院数量激增，在博洛尼亚、乌普萨拉、柏林、圣彼得堡、法国各省会，甚至在北美殖民地的费城，科学院成为民族自豪感和成就的象征。但科学院只是不断发展的科学世界的一种表现形式。与之伴随的是不那么正式的但有时同样重要的社会团体。在巴黎，在皇家科学院创建之前存在的，是在私人住宅或公众场所举办的自然哲学沙龙，有兴趣的人聚集在一起，在一位组织者的领导下讨论、交谈和辩论。它们表明，自然哲学的发展吸引了公众的关注，并且成为一种社会现象。在伦敦，17世纪末兴起的咖啡馆为各类人群会面和讨论问题（包括自然哲学问题）提供了场所。公众的兴趣使得18世纪初出现了公众演示员，这种角色既充当自然哲学家又充当演员，他们用具有异国情调的装置或炫目的陈列品来娱乐和教育公众（要交入场费）。

维系人与人交流的通信网络虽然不像科学院那么显眼，但对于科学史同样重要。自然哲学家们私下交换信件、手稿及其新印制的书籍。信件的私密性使人们可以交流不受欢迎的、激进的全新想法，从而在17世纪的整个欧洲造就了一场基本上秘密进行的讨论。这个无形的“书信共和国”（文艺复兴时期人文主义者的一则习语）将跨越国界、语言和信仰的志同道合的思想家团结在一起，拉近了他们之间的距离。巩固这些通信网络的人被称为“情报员”。他们接收信件，组织和汇编信息，将其分发给有兴趣的各方，并发出后续调查。一位忙碌的情报员的通信量可能非常惊人。曾经鼓励过伽桑狄并且在法国传播伽利略思想的佩雷斯克（1580—1637）一直与约500人保持通信，留下了万余封信件。其中一位通信者米尼姆会托钵修士马兰·梅森（1588—1648）本人就是一个通讯枢纽。在其巴黎的修道院密室中，他接收信件，通过遍布欧洲的网络传播笛卡尔、伽利略等人的工作。在英格兰，三十年战争期间的普鲁士难民塞缪尔·哈特利布（约1600—1662）维持着整个新教欧洲与北美的通信，他幸存下来的2000封信仅仅是他所写全部信件中的一小部分。哈特利布的动力来自依照培根模式对教育、农业和工业进行改革的乌托邦式的实用思想，但也来自宗教信仰，尤其是在英格兰创建一个新教的"人间天堂"的千禧年希望。他的圈子包括企业家、道德家、自然哲学家、神学家和工程师，他的计划从开办技术学院到改进酿酒，可以说应有尽有。科学院本身构成了这个书信网络的节点，而学术期刊——《哲学会刊》、《学者杂志》及其现代后裔——则可被视为这个书信网络定型为墨字的版本。

由于科学院的建立以及技术应用在17世纪的重要性日益增加，在接下来几个世纪，科学工作逐渐专业化，"业余的"自然哲学家渐渐消失。由于日益需要一些知识渊博、值得信赖的人用科学知识和方法来解决实际问题，大学必须用更加正式和严格的训练来培养这些人，这又导致了思想和进路的进一步标准化。由此累积的结果便是19世纪出现了作为一种职业的"科学"，作为一个独特的社会和职业阶层的"科学家"（在某些方面类似于培根在《新大西岛》中的描述），近代早期的世界也逐步演变为现代的科学技术世界。这一转变是一个缓慢而复杂的过程，论述它超出了本书的范围。历史人物选择的道路，影响其决定的想法和需求，使其意图实现或落空的事件，这些既非显而易见亦非注定。虽然现实的自然世界没有什么不同，但人类表达、理解和利用自然界的方式可能非常不同。我们选择的特定历史道路把我们带入了一个充满奇迹的科学技术世界，它会使最伟大的“自然魔法”倡导者感到惊讶，但也并非没有问题，其中既有尚未解决的问题，也有我们自己制造的问题。在一派令人羡慕的自然知识当中，那个智慧、平和、秩序井然的本色列岛继续躲避着我们，即使它一向给我们以启示。

尾声

近代早期自然哲学家留给我们的几乎所有文本和人工制品都表明，他们在热情地探索、创造、保存、测量、收集、组织和学习。他们的无数理论、解释和世界体系在争相寻求认可和接受的过程中命运各异。许多近代早期的概念和发现——哥白尼的日心说、哈维的血液循环理论、牛顿的万有引力平方反比定律——构成了我们现代对世界的理解的基础。其他想法，比如原子论观念和对宇宙尺寸的估计，被后续的科学工作大大更新和完善，而有些想法，比如笛卡尔的旋涡或对磁吸引的机械论解释，则已经完全被抛弃。

现代科学继续探索着近代早期自然哲学家的许多问题和目标——其中一些是他们从中世纪甚至是古人那里继承的。和伽桑狄、笛卡尔和海尔蒙特一样，现代物理学家继续寻求最终的物质粒子，试图理解这些不可见的宇宙单元是如何结合在一起并相互作用而形成世界的。和开普勒、卡西尼和里乔利一样，现代天文学家继续扫视和绘制着天空，用远比第谷、伽利略和赫维留的象限仪和望远镜多样化和功能更强大的的仪器寻找着新的天体和现象。有科学家继承了埃尔南德斯和达·科斯塔等新西班牙探险者的衣钵，继续在丛林和沙漠的动植物中寻找新的药物，或者在深黑的海沟甚至是遥远的世界中寻找新的生命形式。和他们的帕拉塞尔苏斯主义和制金者先辈一样，现代化学家努力修正和改进天然物质，创造新的材料，继续本着玻意耳的精神来认识材料的变化，本着培根的精神来提供对人类生活有用的东西。和维萨留斯、马尔比基和列文虎克一样，现代生物学家和医生用新的仪器来研究动物和人的身体，揭示出更为精细的结构和更令人惊讶的机制。市场上出现的每一种新的电子小发明都反映了技术与奇迹和魔法世界之间的联系。

除了这些连续的环节，还有许多东西发生了变化。促使近代早期自然哲学家研究自然之书——寻找造物主在受造世界中的反映——的深刻的宗教信仰动机，不再是科学研究的主要驱动力。那种恒常不变的历史认识，即认为自己属于一种长期累积的探究自然的传统，已经在很大程度上不复存在了。今天，很少有科学家会像开普勒那样把支持哥白尼学说的教科书冠以"亚里士多德补遗"的副标题，或者像牛顿寻求引力原因那样在古代文献中寻找答案。由于放弃了意义和目的问题，缩小了视野和目标，拘泥于字面意义因而无法理解对于近代早期思想来说如此根本的类比和隐喻，那种内在紧密关联的宇宙图景已经彻底瓦解。具有宽广的思想、活动、经验和专门技能的自然哲学家已经被专业化、专门化的技术科学家所取代。结果导致了一个与更广阔的人类文化和生存视野分离的科学领域。虽然我们必须承认，现代科学技术的发展已经使物质财富和思想成果达到了惊人的水平，但我们无法不认为自己因为丧失了近代早期那种全面的眼界而变得更加可怜。

科学革命时期夹杂着连续和变革，交织着创新和传统。近代早期自然哲学家来自欧洲各地、各个宗教派别、各种社会背景，既有煽动性的创新者，也有谨慎的传统主义者。这些不同角色共同致力于建立对于今天的整个科学世界至关重要的知识体系、机构和方法，这个科学世界与每一个人息息相关。我们可以讲述很多他们极度渴望知道的东西，他们或许也会讲述我们极度渴望听到的东西。对我们而言，他们所处的时代既熟悉又陌生，既像我们自己的时代，又有着显著的不同。近代早期的这种复杂性和热情洋溢使之成为整个科学史上最令人着迷和最重要的时期。

索引

（条目后的数字为原文页码）

A

Academia naturae curiosorum 自然奥秘学院 124
Académie Royale des Sciences 皇家科学院 100, 127–129
Accademia dei Lince 猞猁学院 54, 123
Accademia del Cimento 西芒托学院 123–124
Acosta, José da 阿科斯塔 110
Acquapendente, Girolamo Fabrizio d' 法布里齐奥·阿奎彭登特 101
action at a distance 超距作用 31, 88, 90
Adam's Fall 亚当的堕落 108, 121
Agricola, Georgius 阿格里科拉 115
air 空气 75–77, 100
air-pump 空气泵 77–80
aither 以太 40
Albert the Great, Saint 大阿尔伯特 8
Alchemy 炼金术 80–86 (*see also* chemistry) (*也见化学*)
Aldrovandi, Ulisse 阿尔德罗万迪 112
al-Hazen 阿尔哈增 *see* Ibn al-Haytham *参见伊本·海塞姆*
America 美洲 19, 109–110, 131
discovery of 的发现 15–16
analogy 类比 30, 35, 71, 98, 134
anatomy 解剖学 68, 98–101, 124
theatres 解剖室 99–100
anima motrix 致动灵魂 58, 64
antimony 锑 98
apothecaries 药剂师 85, 97
Arabic learning 阿拉伯学问 6–7, 47, 93
Archeus 阿契厄斯 106–107
Archimedes 阿基米德 12, 114
Aristarchus of Samos 阿里斯塔克 51
Aristotelianism 亚里士多德主义 28–29, 67, 90–92
Aristotelians 亚里士多德主义者 77, 86, 88
Aristotle 亚里士多德 7, 11, 18, 22, 31, 39–40, 43, 58, 64, 72, 84, 86, 101
four causes 四因 24
four elements 四元素 40, 94, 105
on comets 论彗星 55
on varieties of soul 论各种灵 104–105
artillery 火炮 74, 114
artisans 工匠 81, 84, 126
aspects (astrological) 星位（占星学中的）53
astrology 占星学 52–55
medical 医学的 23, 94–97
astronomy 天文学 39–66, 124
atheism 无神论 36, 86, 90

atomism 原子论 9, 85–87, 103, 107, 133
Augustine, Saint 圣奥古斯丁 37, 68
Avicenna 阿维森纳 *see* Ibn Sīnā 参见伊本·西纳

B

Bacon, Sir Francis 弗朗西斯·培根 120–121, 125, 127, 131, 134
bacteria 细菌 102
barometer 气压计 75–76, 124
Bensalem 本色列 121, 132
Berti, Gasparo 贝尔蒂 74, 76
Bessarion, Basilios 贝萨里翁 10
bestiaries 动物寓言集 108
Bible《圣经》37, 68, 105
biblical interpretation《圣经》解释 18, 50, 51, 61–62
biomechanics 生物力学 103
Biondo, Flavio 比翁多 9
black bile 黑胆汁 33, 94–95
blood 血液 94–95, 100–103
Boccaccio 薄伽丘 8
Boerhaave, Herman 布尔哈夫 107
Bologna 博洛尼亚 47, 99, 110, 129
Book of Nature 自然之书 37, 57, 73 (*see also* 'Two Books') *也见*"两本大书"
books of secrets 秘密之书 33
Borelli, Giovanni Alfonso 博雷利 103, 124
botanical gardens 植物园 85, 110, 129 (*see also* Jardin du Roi) *也见*国王花园
botany 植物学 109–111
Botticelli Sandro 波提切利 5
Boyle, Robert 玻意耳 25, 37, 78, 89, 100, 107, 125, 126, 134
Bracciolini, Poggio 布拉乔利尼 9
Brahe, Tycho 第谷·布拉赫 55–56, 58, 65, 134
brain 大脑 94
Brunelleschi, Filippo 布鲁内莱斯基 114
Brunfels, Otto 布伦费尔斯 109
Bruni, Leonardo 布鲁尼 9
Burnet, Thomas 伯内特 69

C

Cabeo, Niccolò 卡贝奥 91
cabinets of curiosities 珍奇馆 111
calendar 历法 49, 61
Calvinism 加尔文主义 85
Cambridge 剑桥 64
Campanella, Tommaso 康帕内拉 34
Campani, Giuseppe 康帕尼 63
Campanus of Novara 诺瓦拉的康帕努斯 50
capillaries 毛细管 101, 102

cartography 地图制作 17, 117, 127
lunar 月球的 63
Casa de Contratación 商局 16
Cassini, Gian Domenico 卡西尼 63, 119, 127, 129, 133
Castelli, Benedetto 卡斯泰利 74
cathedral schools 大教堂学校 5
causal knowledge 因果知识 24, 73
cells 修道院单人小室 102
Cesi, Federico 切西 102, 123
Charles II 查理二世 125
chemiatria 化学医学 *see* chemical medicine *参见*化学医学
chemical medicine 化学医学 85, 89, 90, 123
chemistry 化学, 80–86, 89, 105, 124
China 中国 14, 19, 123, 129
Christ 基督 66, 108
Christianity 基督教 36
Chrysoloras, Manuel 克利索罗拉斯 10
chrysopoeia 制金 80, 83, 84, 89–90
chymical worldview 化学论世界观 83
chymistry, use of the term 对术语化学的使用 80 (*see* chemistry) (*参见*化学)
Cicero 西塞罗 9
Clavius, Christoph 克拉维乌斯 61
Clement VII, Pope 教皇克雷芒七世 49
clocks 时钟 118–119
Colbert, Jean-Baptiste 科尔贝 127
Collegio Romano 罗马学院 61, 112
Columbus, Christopher 哥伦布 15
comets 彗星 1–2, 55–56, 61, 126
commerce 商业 80, 84, 89
compass 罗盘 71, 120
complexion (medical) 体质(医学中的) 53, 96–97, 106
Copernican system 哥白尼体系 59–60, 71
Copernicus, Nicholas 哥白尼 47–51, 63, 133
copper 铜 81, 115
correspondence networks 通信网络 19, 125, 130
cosmographers 宇宙志学者 119
courts, princely 皇宫 122

D

da Costa, Cristóvao 达·科斯塔 110, 134
da Gama Vasco 达·伽马 14
Dante 但丁 8, 85
Dee John 约翰·迪伊 54
deferent 均轮 43, 45, 47
Democritus 德谟克利特 86
Descartes, René 笛卡尔 62–63, 64, 65, 88, 89, 91, 120, 129, 133

determinism 决定论 54, 90
Dionysius the Areopagite (pseudo-) 伪狄奥尼修斯 22
Dioscorides 迪奥斯科里季斯 109
dissection 解剖 68, 98–100, 125, 128
Dominicans 多明我会 15, 34

E

Earth 地球 64, 68–72
age of 的年龄 68
motion of 的运动 59, 61–62
rotation of 的旋转 47, 50, 71
shape of 的形状 15, 129
soul of 的灵魂 71
eccentrics 偏心圆 42–44, 50
ecology 生态学 38
effluvia 流溢 87, 89
elements, four Aristotelian 亚里士多德的四元素 40, 59, 67, 94, 105
fifth 第五元素 40, 67
Elizabeth I 伊丽莎白一世 54, 71, 121
ellipses 椭圆 58, 64
emblematic tradition 象征传统 108–109
Emerald Tablet 《翠玉录》 23
engineers and engineering 工程师与工程 12, 73–74, 114
environmental sciences 环境科学 38, 117
Epicurus 伊壁鸠鲁 86
epicycles 本轮 43–47, 50, 51
epigenesis 渐成论 102–103
Ercker, Lazar 埃尔克 117
Euclid 欧几里得 7
Eudoxus 欧多克斯 42
experiment 实验 71, 73, 77, 84, 86, 104–105, 120, 123
Experimental Philosophy Club 实验哲学俱乐部 125
exploration 探索 14–17

F

Ferdinando II (King of Spain) 费迪南多二世（西班牙国王） 16
Ficino, Marsilio 菲奇诺 10, 22–23, 33–34
fire 火 80, 83, 85, 89, 106
fluid dynamics 流体动力学 74
Fontana, Domenico 丰塔纳 115–116
Fra Angelico 安吉利科 5
Francesca, Piero della 弗朗切斯卡 5
Franciscans 方济各会 15, 82, 110
free will 自由意志 54 (*see also* determinism) (*也见*决定论)
Frontinus 弗龙蒂努斯 9, 12
Fuchs, Leonhart 富克斯 109

G

Galen 盖伦 7, 11, 93, 98–99, 100, 106
Galilei, Galileo 伽利略 59–62, 72–74, 87, 91, 101, 114, 118, 119, 123, 130
gas 气体 106
Gassendi, Pierre 伽桑狄 63, 86–87, 91, 107, 130, 133
Geber 盖伯 85
Gemistos, Georgios 盖弥斯托斯 10
generation 生成 102–104
  spontaneous 自发的 103–104
geocentrism 地心说 42–43, 50, 61
geoheliocentrism 地日心说 *see* Tychonic system 参见第谷体系
geology 地质学 68–70
Gessner, Conrad 格斯纳 108
Gilbert, William 吉尔伯特 58, 70–72
God 上帝 19, 21, 22–23, 24–26, 57, 66, 83–84, 87, 103, 105, 124
  as chymist 作为化学家的 83–84
  as geometer 作为几何学家的 42, 57
  as watchmaker 作为钟表匠的 88, 90
  as romance writer 作为小说作者的 25
gold 金 31, 80, 81, 85, 89, 115
Grant, Edward 爱德华·格兰特 8
Grassi, Orazio 格拉西 61
gravitation 引力 64–66, 133
Great Chain of Being 伟大的存在之链 23
Greek learning 希腊学问 10
Gregory XIII, Pope 教皇格里高利十三世 61
Gresham College 格雷欣学院 125
Grew, Nehemiah 格鲁 103
Grimaldi, Francesco Maria 格里马尔迪 63, 91
Guarino da Verona 瓜里诺·达·维罗纳 10
gunpowder 火药 69, 114, 117, 120
Gutenberg, Johannes 古腾堡 13

H

Halley, Edmond 哈雷 69
Harrison, John 哈里森 119
Hartlib, Samuel 哈特利布 131
Hartmann, Johannes 哈特曼 85
Harvey, William 哈维 100–101, 102, 103, 133
heart 心脏 31 100–101
heliocentrism 日心说 47–52, 57, 61–63
heliotropism 向日性 28, 34
Hell 地狱 70, 85
Helmontianism 海尔蒙特理论 98 105–107
Henry of Langenstein 朗根施泰因的亨利 8
Henry the Navigator 航海家亨利 14
herbals 本草志 108

Hermes Trismegestus 三重伟大的赫尔墨斯 22–23

Hermetica 《赫尔墨斯文集》23 66

Hernández, Francisco 埃尔南德斯 110, 123, 134

Hero 希罗 12

Hevelius, Johann 赫维留 63, 134

Hicetas 希克塔斯 51

Hippocrates 希波克拉底 93, 97, 107

Hooke, Robert 胡克 65, 78, 102, 118, 125

houses (astrological) 宫（占星学中的） 53

humanism 人文主义 8–12, 18, 51, 66, 109, 114, 122, 130

'secular' "世俗的" 11

humoral theory 体液理论 33, 53, 94–95, 106–107

Huygens Christiaan 惠更斯 63, 118, 127

hydraulics 水力学 9, 74, 103

I

Iamblichus 扬布里柯 34

Ibn al-Haytham 伊本·海塞姆 46–47

Ibn Sīnā 伊本·西纳 93

India 印度 14, 19, 110

inertia 惯性 64, 72

Inquisition 宗教裁判所 62

insects 昆虫 104

intelligencers 情报员 130–131

iron 铁 81, 85, 89, 115

Islamic science 伊斯兰科学 6–7

italics 斜体 13

J

James I 詹姆斯一世 121

Jardin du Roi 国王花园 85, 128

Jean of Rupescissa 鲁庇西萨的约翰 82

Jesuits 耶稣会士 19, 59–63, 70, 91, 105, 110, 129

Jonson Ben 琼森 85

Journal des sçavans 《学者杂志》 131

Jupiter 木星 34, 47

moons of 的卫星 59, 60, 119

K

Kepler Johannes 开普勒 56–58, 63, 64, 72, 133, 134

kinematics 运动学 72–74 91 (*see also* motion) *也见*运动

Kircher Athanasius 基歇尔 25, 28, 31, 69–70, 74, 112

L

law 法学 93
lead 铅 81, 88
Leibniz Gottfried Wilhelm 莱布尼茨 65
Leiden 莱顿 99, 107, 110
Lemery Nicolas 莱默里 85
Leonardo da Vinci 达·芬奇 5, 114
Leopold I Emperor 皇帝利奥波德一世 124
Leopoldina 德国科学院 124
Leucippus 留基伯 86
light 光 52, 77, 87, 119, 126
lions 狮子 31, 100
longitude 经度 117–119, 127, 129
Louis XIV 路易十四 63, 98, 100, 127
Louvain 鲁汶 105
Lucretius 卢克莱修 9, 86
Luther Martin 马丁·路德 17–18

M

macrocosm-microcosm 大宇宙—小宇宙 23, 30–31, 36, 94, 101, 112
Magdeburg Sphere 马格德堡半球 77
magia naturalis 自然魔法 27–35, 65, 89, 120, 132
magnetism 磁学 25, 29, 64, 71–72, 89, 91
Malpighi Marcello 马尔比基 101, 124, 126, 134
Manilius 马尼留斯 9
Manutius Aldus 马努提乌斯 13
Mars 火星 45, 47, 54, 58
materialism 唯物论 90
mathematics 数学 11, 19, 41, 60, 65, 73, 91
mechanical philosophy 机械论哲学 65, 87–90, 102, 107
Medici 美第奇 68, 103
  Cosimo I de' 科西莫一世·德· 10
  Cosimo II de' 科西莫二世·德·59
  Ferdinando II de' 费迪南多二世·德· 75
  Leopoldo de' 莱奥波尔多·德· 123
medicine 医学 32–33, 53–54, 82, 89, 93–98, 105, 106–107, 109, 110, 124
medical education 医学教育 99, 107, 110
Medieval Warm Period 中世纪暖期 6, 8
melancholy 忧郁 33
Melanchthon Philipp 梅兰希顿 18, 49, 55
Mercator Gerhardus 墨卡托 117
Mercury (chymical principle) 汞（化学要素） 81, 83
Mersenne Marin 马林·梅森130
metals, formation of 金属的形成 81
meteors 流星 55

科学革命

microscope 显微镜 101–102
mining 采矿 84, 115–117
Monardes Nicholás 莫纳德斯 110
Montpellier 蒙彼利埃 98, 110
Moon 月亮 41, 47, 53, 94
mapping of 绘制月面图 63
Moritz of Hessen-Kassel 莫里茨 85
motion 运动 72–74, 91
circular 圆周的 42, 67, 72, 101
of falling bodies 落体的 43, 72
natural 自然的 50, 67
planetary 行星的 58, 64–65
projectile 抛射体的 74
museums 博物馆 112, 122

N

natural ‘magic’ 自然“魔法” *see magia naturalis* 参见自然魔法
natural history 博物志 109–111, 120, 127
natural philosophy, defined 经过界定的自然哲学 27
natural place 自然位置 50, 67, 71–72
navigation 导航 17, 117–118
Neoplatonism 新柏拉图主义 22–23, 34
Newton Sir Isaac 牛顿 64–66, 72, 90, 107, 126–127, 133
Nicholas V Pope 教皇尼古拉五世 11

O

oak galls 橡树瘿 104
observation 观察（测） 29, 73, 120
observatories of Paris 巴黎天文台 129
of Greenwich 格林威治的 129
oceanic currents 洋流 70
Oldenburg Henry 奥尔登堡 125
optics 光学 33, 126
ores 矿石 81, 117
Oresme Nicole 奥雷姆 8, 50
Orta Garcia de 德·奥塔 110
Osiander Andreas 奥西安德尔 49, 51
Oxford 牛津 100, 110, 125
Oxford Calculators 牛津计算者 73

P

Padua 帕多瓦 74, 99, 100, 110
Paracelsus Theophrastus von 帕拉塞尔苏斯
Hohenheim and Paracelsianism 帕拉塞尔苏斯与帕拉塞尔苏斯主义 82–84, 98, 105, 106
parallax 视差 50, 55
Pascal Blaise 帕斯卡 75–77
patronage 赞助 68, 122, 124, 126, 127, 129
Paul III Pope 教皇保罗三世 51
Peiresc Nicolas-Claude Fabri de 佩

雷斯克 130
pendula 摆 72, 118
Périer Florin 佩里耶 75–76
Perrault Claude 佩罗 100
Petrarch 彼特拉克 8
Peurbach Georg 普尔巴赫 46
pharmacy 药学 82–83, 85, 89, 97
Philip II 菲利普二世 110
Philosophers' Stone 哲人石 81, 82, 89
Philosophical Transactions 《哲学会刊》 126, 131
phlegm 黏液 94–95
physics 物理学 42–43, 65, 71–74
Pierre de Maricourt 马里古的皮埃尔 71
Pisa 比萨 110
Pius II Pope 教皇庇护二世 11
plague 瘟疫 8, 106–107
planetary distances 行星距离 50, 56–57
planets 行星 41, 53
  motion of 的运动 41–43, 45, 51, 58
  number of 的数量 56–57
  retrograde 的逆行 41, 45, 47–48
Plato 柏拉图 10, 11, 32, 41–42, 73
Platonic solids 柏拉图正多面体 56–57
plenum 实满 77
Pletho 普莱东 *see* Gemistos Georgios *参见*盖弥斯托斯
Pliny the Elder 老普林尼 108, 109–110
Plotinus 普罗提诺 34
pole star 北极星 40
Porta Giambattista della 德拉·波塔 32, 123
portents 征兆 54–55
preformationism 预成论 102–103
printing 印刷术 12–14, 53, 99, 120
prisca sapientia 古代智慧 66
professionalisation 专业化 122, 131
prophecies 预言 66
Protestantism 新教 17–18, 62
Ptolemy Claudius 托勒密 7, 15, 42–44, 52, 117
Puy-de-Dome experiment 多姆山实验 77
Pythagoreans 毕达哥拉斯主义者 11, 41

Q

qualities 性质 33–34, 73, 87, 94
  manifest and hidden 明显的与隐秘的 28–29, 65, 88
  primary 第一性质 28–29, 87
  secondary 第二性质 87
quintessence 第五元素 40, 59, 67
Quintilian 昆体良 9

R

Redi Francesco 雷迪 104, 124
Reformations Protestant and Catholic 新教与天主教的宗教改革 17–19
religious motivations for science 科学的宗教动机 19, 36–37, 57, 134
Renaissance Carolingian 加洛林王朝的文艺复兴 5, 9
Italian 意大利的 5, 8
of the 12th century 12世纪的 6
resurrection 复活 83
Reuchlin Johannes 罗伊希林 18
revelation 启示 37
Rheticus Georg Joachim 雷蒂库斯 49, 51–52
Riccioli Giovanni Battista 里乔利 60, 91, 133
Roemer Ole 罗默 119
Royal College of Physicians 皇家医师学院 98
Royal Society of London 伦敦皇家学会 64–65, 100, 125–127, 129
Rudolf II 鲁道夫二世 58

S

Sacrobosco 萨克罗博斯科 46, 55
Sahagún Bernardino de 萨阿贡 110
Saint Peter' s Basilica 圣彼得教堂 54, 115
salons 沙龙 130
Salt (chymical principle) 盐（化学要素） 83
saltpetre 硝石 89, 117
Saturn 土星 33–34, 47, 50
rings of 土量 59, 60, 63
scala naturae 自然阶梯 22–23
Scheiner Christoph 沙伊纳 61
Scholasticism 经院哲学 7, 11, 18, 90 (*see also* Aristotelianism) *也见*亚里士多德主义
Schönberg Nicolaus 舍恩贝格 49
Schreck Johann 施莱克 123
science and religion 科学与宗教 36–37, 61–62 (*see also* theology *and* God) （*也见*科学*和*神）
scientific journals 科学期刊 131
scientific societies 科学社团 100, 122–129
seasons 四季 42, 44, 95
secrecy 保密 81
secular humanism 世俗人文主义 *see* humanism *参见*人文主义
semen 精液 102
semina ('seeds') 种子 106
sense perception 感官知觉 28–29, 87 104
Shadwell Thomas 沙德韦尔 126

索引

sign (zodiacal) 宫（黄道带） 41, 53, 94
signatures 征象 29–30
silver 银 81, 86, 115
simple harmonic motion 简谐运动 118
Sixtus IV Pope 教皇西克斯图斯四世 11
Sixtus V Pope 教皇西克斯图斯五世 115
Solomon' s House 所罗门宫 121, 122, 125
souls 灵魂 102, 104–105
human 人的 83, 87
world 世界 32
sound 声音 77
spagyria 炼金术 83–84, 90
spheres celestial 天球 42–43, 46, 56, 58, 64
spirits, animal or vital 动物精气或生命精气 32, 103, 107
of the world 世界精气 31–32
Stahl Georg Ernst 施塔尔 107
stars 星 40, 50–51
Medicean 美第奇 59
Ludovican 卢多维奇 63
Stelluti Francesco 斯泰卢蒂 102
Steno Nicholas the Blessed 斯泰诺 *see* Stensen, Niels 参见尼尔斯·斯滕森
Stensen Niels 尼尔斯·斯滕森 68, 123
Strabo 斯特拉波 10
strata 地层 68
sublunar/superlunar worlds, defined 经过界定的月下/月上世界 25, 39–40, 67
Sulphur (chymical principle) 硫（化学要素） 81, 83
Sun 太阳 30–31, 34, 40, 44, 46, 58
sunflowers 向日葵 28
sunspots 太阳黑子 61, 123
supernova 超新星 55
superstition 迷信 35
surgery 外科手术 97
Swift Jonathan 斯威夫特 126
sympathies 共感 31, 34–35, 65, 72, 88

T

Tartaglia Niccolò 塔尔塔利亚 74, 114
taxonomy 分类学 110
technology 技术 113–121
telescope 望远镜 59, 63, 122, 126
temperament 气质 94
Thales 泰勒斯 105
Theodoric of Freiburg 弗赖贝格的狄奥多里克 8
theology 神学 8, 22, 25, 26, 36, 57, 66, 88, 93, 104, 105
Theophrastus 帕拉塞尔苏斯 91
thermometer 温度计 124

Thirty Years War 三十年战争 62, 131
tides 潮汐 35, 53, 61
Timaeus 《蒂迈欧篇》 32
tin 锡 81
Titian 提香 99
Torricelli Evangelista 托里拆利 75
transfusions 输血 101
translation movement 翻译运动 7
transmutation 嬗变 80–81, 84, 89
Trent council of 特伦托会议 18
tria prima 三要素 83–84, 105
'Two Books' "两本大书" 37, 57
Tychonic system 第谷体系 56, 59–60, 63

U

universities 大学 6, 7–8, 11, 18, 64, 84, 93, 122, 131
Uraniborg 天堡 56
Urban VIII Pope 教皇乌尔班八世 34, 62, 102

V

vacuum 真空 77–78
van Helmont Joan Baptista 海尔蒙特 98, 105–107, 133
van Leeuwenhoek Antoni 列文虎克 102, 126, 134
Venus (phases) 金星（位相） 59, 60
Vesalius Andreas 维萨留斯 99, 109, 134
Vespucci Amerigo 亚美利哥·韦斯普奇 16
Vesuvius 维苏威火山 69
vitalism 活力论 102, 104, 107
Vitruvius 维特鲁威 9, 12, 114
volcanoes 火山 69–70, 113
von Guericke Otto 冯·盖里克 77
vortices 涡旋 64, 65, 133

W

water 水 74, 105–106, 114
weather 天气 53
Whiston William 惠斯顿 69
Widmannstetter Johann Albrecht 维德曼施泰特 49
willow tree experiment 柳树实验 106
wine 酒 82, 98
Worm, Ole 沃姆 111
Wren, Christopher 雷恩 125

Y

yellow bile 黄胆汁 94–95

Z

zodiac 黄道带 41, 47

zoology 动物学 109–111

Lawrence M. Principe

THE SCIENTIFIC REVOLUTION

A Very Short Introduction

Contents

Acknowledgements xi

List of illustrations xiii

Introduction 1

1 New worlds and old worlds 4

2 The connected world 21

3 The superlunar world 39

4 The sublunar world 67

5 The microcosm and the living world 93

6 Building a world of science 113

Epilogue 133

References 136

Further reading 137

Acknowledgements

I would like to thank my friends and colleagues who have read and critiqued all or part of the manuscript of this book, or with whom I have discussed and whined about the challenge of compressing the Scientific Revolution into so compact a form, especially Patrick J. Boner, H. Floris Cohen, K. D. Kuntz, Margaret J. Osler, Gianna Pomata, María Portuondo, Michael Shank, and James Voelkel. I would also like to thank the people who have provided the images for this book: James Voelkel at the Chemical Heritage Foundation; Earle Havens and his colleagues in Special Collections at the Sheridan Libraries, Johns Hopkins University; and David W. Corson and his colleagues at the Rare and Manuscript Collections, Kroch Library, Cornell University.

In particular, I wish to remember here the many conversations held over single-malt Scotch with my colleague and friend Maggie Osler about how to write the history of early modern science. Her premature death leaves the world a poorer and less mischievous place; I dedicate this volume to her memory.

List of illustrations

1 The interconnected world **26**
Courtesy of the Division of Rare and Manuscript Collections, Cornell University Library

2 Aristotle's universe **43**

3 Ptolemy's eccentric **44**

4 Ptolemy's epicycles and deferents **45**

5 Ibn al-Haytham's thick-spheres planetary model **46**
Courtesy of the Johns Hopkins University, The Sheridan Libraries, Rare Books and Manuscripts Department

6 Copernicus's explanation of retrograde motion **48**

7 Three world systems compared emblematically **60**
Courtesy of the Division of Rare and Manuscript Collections, Cornell University Library

8 Kircher's view of the Earth's interior **70**
Courtesy of the Roy G. Neville Historical Chemical Library, Chemical Heritage Foundation, Philadelphia

9 Berti's water barometer and Torrecelli's mercury barometer **76**
Courtesy of the Roy G. Neville Historical Chemical Library, Chemical Heritage Foundation, Philadelphia

10 Von Guericke's Magdeburg spheres **78**
Courtesy of the Roy G. Neville Historical Chemical Library, Chemical Heritage Foundation, Philadelphia

11 Boyle's and Hooke's air pump **79**
Courtesy of the Roy G. Neville Historical Chemical Library, Chemical Heritage Foundation, Philadelphia

12 Alchemical allegory of purification **82**
Courtesy of the Roy G. Neville Historical Chemical Library, Chemical Heritage Foundation, Philadelphia

13 Square of the elements and humours **95**

14 Bodily organs and zodiacal correspondences **96**
Courtesy of the Roy G. Neville Historical Chemical Library, Chemical Heritage Foundation, Philadelphia

15 Worm's cabinet of curiosities **111**
Courtesy of the Roy G. Neville Historical Chemical Library, Chemical Heritage Foundation, Philadelphia

16 Moving the Vatican obelisk **116**
Courtesy of the Johns Hopkins University, The Sheridan Libraries, Rare Books and Manuscripts Department

17 A dissection at the Royal Academy of Sciences **128**
Collection of the author

Introduction

Late in 1664, a brilliant comet appeared in the skies. Spanish observers were the first to note its arrival, but over the following weeks, as it grew in size and brightness, eyes all over Europe turned towards this heavenly spectacle. In Italy, France, Germany, England, the Netherlands, and elsewhere – even in Europe's young colonies and outposts in the Americas and Asia – observers tracked and recorded the comet's motions and changes. Some took careful measurements and argued over calculations of the comet's size and distance, and whether its path through the heavens was curved or straight. Some observed it with the naked eye, others with instruments such as the telescope, an invention then just about sixty years old. Some tried to predict its effects on the Earth, on the weather, on the quality of the air, on human health, and on the affairs of men and the fates of states. Some saw it as an opportunity to test new astronomical ideas, others saw it as a divine portent for good or ill, and many saw it as both. Pamphlets flowed from printing presses, articles and contentions appeared in the new periodicals devoted to natural phenomena, people discussed it in princely courts and academies, in coffee-houses and taverns, while letters full of ideas and data shuttled back and forth among distant observers, weaving webs of communication across political and confessional boundaries. All of Europe watched this spectacle of nature and strove to understand it and to learn from it.

The comet of 1664–5 provides but one instance of the ways in which 17th-century Europeans paid close attention to the natural world around them, interacted with it and with each other. Peering through ever-improving telescopes, they saw immense new worlds – undreamt-of moons around Jupiter, the rings of Saturn, and countless new stars. With the equally new microscope, they saw the delicate details of a bee's stinger, fleas enlarged to the size of dogs, and discovered unimagined swarms of 'little animals' in vinegar, blood, water, and semen. With scalpels, they revealed the internal workings of plants, animals, and themselves; with fire, they analysed natural materials into their chemical components, and combined known substances into new ones. With ships, they sailed to new lands, and brought back amazing reports and samples of novel plants, animals, minerals, and peoples. They devised new systems to explain and organize the world and revived ancient ones, ceaselessly debating the merits of each. They sought for causes, meanings, and messages hidden in the world, for the traces of God's creative and sustaining hand, and for ways to control, improve, and exploit the worlds they encountered with both new technology and hidden ancient knowledge.

The Scientific Revolution – roughly the period from 1500 to 1700 – is the most important and talked-about era in the history of science. Ask ten historians of science about its nature, duration, and impact, and you are likely to get fifteen answers. Some see the Scientific Revolution as a sharp break from the medieval world – a time when we all (Europeans at least) became 'modern'. In this view, the 16th and 17th centuries were truly revolutionary. Others have tried to make the Scientific Revolution into a non-event, a mere illusion of retrospection. More circumspect scholars nowadays, however, recognize the many important continuities between the Middle Ages and the Scientific Revolution, but without denying that the 16th and 17th centuries reworked and built upon their medieval inheritance in significant and stunning ways. Indeed, the 'scientific revolution', now more frequently called the 'early modern period', was a time of *both* continuity and change. It saw a substantial increase in the number of people asking questions about the natural world, a proliferation of new answers to those questions, and the development of new ways of gaining answers. This book describes some of the ways early modern thinkers envisioned and engaged with the worlds around them, what they found in them, and what it all meant for them. It outlines how they laid many of the foundations that continue to undergird modern scientific knowledge and methods, wrestled with questions that continue to trouble us, and even crafted rich worlds of beauty and promise that we have often forgotten how to see.

Chapter 1
New worlds and old worlds

Early modern accomplishments grew upon intellectual and institutional foundations established in the Middle Ages. Many of the questions early moderns strove to answer were posed in the Middle Ages, and many methods used for answering them were products of medieval investigators. Yet early modern scholars loved to disparage the medieval period and to claim that their work was wholly new, despite the fact they retained and relied upon at least as much as they discarded, or retailored it to fit the changing times. Specific changes between the Middle Ages and the early modern period, whether intellectual, technological, social, or political, did not occur simultaneously across Europe. Recognizably 'modern' developments in such areas as medicine, engineering, literature, art, economic and civic affairs were thoroughly established in Italy well before they appeared in more peripheral parts of Europe like England. Similarly, periods of development occurred at different times and speeds within different scientific disciplines. The period roughly 1500 to 1700 – call it what you will – was a rich tapestry of interwoven ideas and currents, a noisy marketplace of competing systems and concepts, a busy laboratory of experimentation in all areas of thought and practice. Text after text from the period testifies to the excitement their authors felt about their own times. One label, one book, one scholar, one generation will not comprehend it in its totality.

To begin to understand it and its significance, we need to look closely at what actually took place then and why.

Understanding the Scientific Revolution requires understanding first its background in the Middle Ages and Renaissance. In particular, the 15th century witnessed significant changes in European society and a massive broadening of Europe's horizons, both literally and figuratively. Four key events or movements fundamentally reshaped the world for people living in the 16th and 17th centuries: the rise of humanism, the invention of movable-type printing, the discovery of the New World, and the reforms of Christianity. While not strictly scientific developments, these changes reshaped the world for thinkers of the period.

The Renaissance and its medieval origins

The term 'Italian Renaissance' usually brings to mind masterpieces of art and architecture by well-known figures like Sandro Botticelli, Piero della Francesca, Leonardo da Vinci, Fra Angelico, and many others. But the Renaissance saw much more than a blossoming of fine arts. Literature, poetry, science, engineering, civic affairs, theology, medicine, and other fields prospered as well. The brilliance and importance of the 15th-century Italian Renaissance for history and for modern culture should not be underestimated. All the same, it should also be remembered that it was not the first significant flowering of European culture after the 5th-century collapse of Classical civilization that followed the fall of the Roman Empire. There had been at least two earlier 'renaissances' (a word which means 'rebirth').

The first, the Carolingian Renaissance, followed the late 8th-century military campaigns of Charlemagne that brought greater stability to Central Europe for much of the 9th century. Charlemagne's court at Aachen (Aix-la-Chapelle) became a centre of learning and culture. The cathedral schools that would later provide the foundations for universities trace their origins to this period. Charlemagne's crowning by Pope Leo III in 800 as 'Emperor of the Romans' encapsulates a basic theme of Carolingian reforms: the attempt to return to the glory of ancient Rome. Architecture, coinage, public works, and even writing styles were devised to reproduce the way imperial Romans had done things, or at least the way 9th-century people imagined the Romans had done things. This flowering was, however, short-lived.

The second 'rebirth' of Latin Europe was much broader and more permanent. Its momentum carried forward, although diminished in intensity, to the start of the Italian Renaissance. This second 'rebirth' was the 'Renaissance of the Twelfth Century', a great explosion of creativity in the sciences, technology, theology, music, art, education, architecture, law, and literature. The triggers for this efflorescence remain open to debate. Some scholars point to a warmer, more favourable climate for Europe beginning in the 11th century (called the 'Medieval Warm Period') coupled with improvements in agriculture that brought enough food and prosperity for Europe's population to double and perhaps triple within a relatively short time. The rise of urban centres, more stable social and political systems, more abundant food, and thus more time for thought and scholarship, all contributed to initiating this Renaissance.

The intellectual appetite of a reawakened Europe found rich fare on which to feed in the Muslim world. As Christian Europe began to push back against the frontiers of Islam in Spain, Sicily, and the Levant, it encountered the wealth of Arabic learning. The Muslim world had become heir to ancient Greek knowledge, translated it into Arabic, and enriched it many times over with new discoveries and ideas. In astronomy, physics, medicine, optics, alchemy, mathematics, and engineering, the *Dār al-Islām* ('Habitation of Islam') towered over the Latin West. Europeans wasted no time in acknowledging this fact, nor in exerting themselves to acquire and assimilate Arabic learning. European scholars embarked upon a great 'translation movement' in the 12th century. Dozens of translators, often monastics, trekked to Arabic libraries, especially in Spain, and churned out Latin versions of hundreds of books. Significantly, the texts they chose to translate were almost entirely in the areas of science, mathematics, medicine, and philosophy.

The Latin Middle Ages had inherited from the Classical world only those texts the Romans possessed; by the end of the empire, only a handful of Roman scholars could read Greek, and therefore virtually the only texts the Romans had to pass on were Latin paraphrases, summaries, and popularizations of Greek learning. It was as if our successors got only newspaper accounts and popularizations of modern science and virtually no scientific journals or texts. Thus scholars of the Latin Middle Ages revered the names of the great authors of antiquity and had descriptions of their ideas, but possessed almost none of their writings.

The 12th-century translators changed all that. They translated works of original Arabic authorship and Arabic translations of ancient Greek works. The majority of ancient Greek texts thus came to Europeans in Arabic dress. From Arabic came the medicine of Galen, the geometry of Euclid, the astronomy of Ptolemy, and virtually the entire corpus of Aristotle we have today – not to mention the more advanced works of Arabic authors in all these fields and more. Around 1200, this explosion of knowledge crystallized into curricula for perhaps the most enduring legacy of the Middle Ages for science and scholarship: the university. Aristotle's writings on natural philosophy formed a core of the curriculum, and his logical works gave rise to Scholasticism, a rigorous and formalized methodology of logical inquiry and debate applicable to any subject, and upon which university studies were based.

The importance of the university as an institutional home for scholarship cannot be overemphasized. As the prominent scholar

Edward Grant writes, the medieval university 'shaped the intellectual life of Western Europe'. While the highest degree in the university was in theology, one could not become a theologian without first mastering the logic, mathematics, and natural philosophy of the day, since those topics were employed routinely in the advanced Christian theology of the Middle Ages. Indeed, most great natural philosophers of the period were doctors of theology: St Albert the Great (now patron saint of natural scientists), Theodoric of Freiburg, Nicole Oresme, Henry of Langenstein. All these figures were educated in, taught in, and found a home in a university.

The vigorous cultural life of the 13th century was checked by the disasters of the 14th. Early in the century – possibly as a result of the end of the Medieval Warm Period – repeated crop failures and famine struck a now overpopulated Europe. At mid-century, the Black Plague swept across Europe with astonishing swiftness, killing its victims within a week of infection. We have no experience today of any loss of life or societal upheaval as rapid, unstoppable, or devastating as the reign of the Black Death. In four years, from 1347 to 1350, it killed roughly half of Europe's population. The first signs of a distinctive Italian Renaissance had begun to appear just before these troubled times – the poet Dante (1265–1321) was active before the plague, while the younger writers Boccaccio (1313–75) and Petrarch (1304–74) lived through it.

Humanism

The Italian Renaissance, fully underway a generation or two after the peak plague years, provided the first key background for the Scientific Revolution: the rise of *humanism*. Humanism proves difficult to define succinctly and rigorously. It is better to speak of *humanisms* – a collection of related intellectual, literary, sociopolitical, artistic, and scientific currents. Among the most widely shared beliefs of humanists was the conviction that they were living in a new era of modernity and novelty, and that this new era was to be measured with respect to the accomplishments of the ancients. They looked for a *renovatio artium et litterarum* (a renewal of arts and letters) to be brought about in part through the study and emulation of ancient Greeks and Romans. Accordingly, it was humanist historians of the Italian Renaissance – such as the Florentines Leonardo Bruni (1369–1444) and Flavio Biondo (1392–1463) – who devised the three-fold periodization of history with which we are all familiar (and from whose implications we still must struggle to free ourselves). According to this periodization, the antiquity of Greece and Rome constitutes the first era, while the third era is that of modernity, beginning of course with the Renaissance authors themselves. Falling between these two high points, according to the humanists, lies a 'middle' period of dullness and stagnation, which is thus called the 'Middle' Ages. Indeed, perhaps the most enduring invention of the Renaissance has been the concept of the Middle Ages, to the extent that we have no name for the period 500 to 1300 that is not suffused with the disdain Italian humanists felt towards it. Given the recent memory of famine and plague years as their immediate background, the restoration of prosperity in Italy around 1400 must surely have seemed the dawn of a 'new age'.

Imitation is supposed to be the sincerest form of flattery, and humanists expressed their admiration of antiquity by imitating Roman styles. Attempts to return to antiquity had happened before, notably in the Carolingian Renaissance 600 years earlier. The grandeur of Rome casts a very long shadow indeed in human memory. The humanist hunger to know more about that past era expressed itself in a quest for long-lost Classical texts. One early humanist, Poggio Bracciolini (1380–1459), taking advantage of recesses during the reform-minded Church Council of Konstanz (1414–18), where he was employed as apostolic secretary, ransacked nearby monastic libraries searching for survivals of Classical literature. He found Quintilian on rhetoric and previously unknown orations of Cicero, but – of greater importance for the history of science – he found also Lucretius' *On the Nature of Things*, a work that presented ancient notions of atomism, Manilius on astronomy, Vitruvius on architecture and engineering, Frontinus on aqueducts and hydraulics. These works had been copied and preserved through the centuries by medieval monks, and had lain – perhaps in just a single surviving copy – in their monastic libraries for generations.

The humanists' recovery of Roman learning was paired with a revival of the study of Greek. The background for the revival of Classical Greek, almost completely unstudied in the Latin West for a thousand years, was the arrival of Greek diplomats and churchmen on embassies to Italy around 1400. Their mission was to secure aid against the Turkish threat and a reunion of Eastern and Western Churches, divided by schism since 1054. One of the first, Manuel Chrysoloras (c. 1355–1415), arrived as a diplomat but stayed as a teacher of Greek; many prominent humanists were his students. Their appetites whetted for Greek texts, Italians travelled to Constantinople to hunt after manuscripts. Guarino da Verona (1374–1460) brought back crates of manuscripts that included Strabo's *Geography*, which he then translated. It is said that one crate of manuscripts was lost in transit, which made Guarino's hair turn grey overnight from grief. The Greek delegation to the Council of Florence in the 1430s included two notable Greek scholars. One was Basilios Bessarion (1403–72), later made a cardinal, who gave his collection of nearly a thousand Greek manuscripts to Venice. The other was a strange character named Georgios Gemistos, known as Pletho (c. 1355–c. 1453), who later advocated a return to ancient Greek polytheism. Pletho taught Greek in Florence, and brought the works of Plato and Platonists to the attention of the West. His teaching led the ruling Duke Cosimo I de' Medici to found a Platonic Academy in Florence. Its first leader, Marsilio Ficino (1433–99), translated the works of Plato and texts by several later Platonists, most of which had been unknown to Western European readers.

Thus the 15th century saw the recovery of huge numbers of ancient texts – many on scientific and technological topics – much as the 12th century had done. But humanists were distinguished not so much by a love of texts, as by a love of *pure and accurate* texts. They disdained the texts of Aristotle and Galen used in universities as corrupt – full of barbarisms, 'Arabisms', accretions, and errors. They rejected Scholasticism as sterile, barbarous, and inelegant. They considered the universities (particularly the northern ones, less so those in Italy) as relics of those stagnant 'Middle' Ages, and chided their scholars for writing a degraded Latin, devoid of elegance. Thus an important feature of humanism was its establishment of new scholarly communities outside the universities.

There is a modern misconception that humanists were somehow secularist, irreligious, or even anti-religious. It is true that some humanists criticized ecclesiastical abuses and disdained Scholastic theology, but in no way whatsoever did they reject Christianity or religion. Indeed, many advocated church reforms parallel with their desired reform of language – by a return to antiquity, to the Church of the first several centuries AD. Many humanists were in Holy Orders, employed in ecclesiastical administration, or supported by church benefices, and the Catholic hierarchy patronized humanism. Many Renaissance-era Popes were fervent humanists – particularly Nicholas V, Sixtus IV, and Pius II – as were their cardinals and courts, where humanists were encouraged. The modern error comes from a confusion with so-called *secular humanism*, an invention of the 20th century that has no counterpart in the early modern period.

Renaissance humanism's impact on the history of science and technology was both positive and negative. On the positive side, humanists made available hundreds of important new texts, and promoted a new level of textual criticism. The reintroduction of Plato, thanks especially to his adoption of Pythagorean mathematics, raised the status of mathematics and provided an alternative to the Aristotelianism favoured at universities. The desire to measure up to the ancients inspired engineering and building projects across Italy, with the ancient engineers Archimedes, Hero, Vitruvius, and Frontinus as models. On the downside, the adulation of antiquity could go too far by rejecting everything after the fall of Rome as barbarism. It is thus that Europe began to lose its respect for and knowledge of Arabic and medieval achievements, which in the sciences, mathematics, and engineering were – let there be no doubt – substantial advancements over the ancient world.

The invention of printing

The invention of movable-type printing around 1450 well served the humanist interest in texts. This invention, or at least its successful deployment, is credited to Johannes Gutenberg (c. 1398–1468), originally a goldsmith in Mainz. The key to movable-type printing was the creation of cast metal type, each bearing a single raised letter. These type could be assembled into full pages of text, their surfaces smeared with an oil-based ink and pressed against paper, thus printing an entire page (or set of pages) at once. After printing numerous copies, the page of type could be taken apart and the letters readily rearranged into the next set of pages. Previously, books had to be copied by hand, resulting in slow production and high price. The late medieval growth of universities and increase in literacy created a demand for books that outstripped the supply, exerting pressure to produce books more quickly, thus leading to book-making enterprises outside the traditional monastic and university scriptoria. This increased production led to more copying errors – something humanists deplored. Printing allowed for faster and more reliable production, although the labour involved in paper-making, typesetting, and printing meant that books remained expensive. (Gutenberg's Bible, printed in 1455, cost 30 florins, more than a year's salary for a skilled workman.)

The transition to print was not immediate. Manuscripts continued to exist alongside books, although their use was increasingly limited to the restricted circulation of private, rare, or privileged materials. Printed typefaces mimicked manuscript writing; in Northern Europe this meant Gothic bookhands, but Italy, Venice in particular, soon became the centre of the printing industry. Italian printers, such as Teobaldo Mannucci, better known by his Latinized humanist name Aldus Manutius (1449–1515), adopted the cleaner, crisper shapes of letters developed by Italian humanists (which they thought imitated the way Romans wrote), thereby creating fonts that not only displaced older ones, but also formed the basis for most fonts used today; hence our elegant slanted font is still known as 'Italic'.

Printing presses sprang up rapidly across Europe. By 1500, there were about a thousand in operation, and between thirty and forty thousand titles had been printed, representing roughly ten million books. This flood of printed material only increased throughout the 16th and 17th centuries. Books became steadily less expensive (often with a loss of quality) and easier for less wealthy buyers to obtain. Printing allowed for faster communication through broadsides, newsletters, pamphlets, periodicals, and a slew of other paper ephemera. Although most of these ephemera perished soon after their production (like last week's newspaper), such items were very common in the early modern period. The press thus created a new world of the printed word – and of literacy – like never before known.

One easily overlooked feature of printing was its ability to reproduce *images and diagrams*. Illustrations posed a problem for the manuscript tradition since the ability to render drawings accurately depended upon the copyist's draftsmanship, and often upon his understanding of the text. Consequently, every copy meant degradation for anatomical renderings, botanical and zoological illustrations, maps, charts, and mathematical or technological diagrams. Some copyists simply omitted difficult graphics. Printing meant that an author could oversee the production of a master woodcut or engraving, which could then produce identical copies easily and reliably. Under such conditions, authors were more willing and able to include images in their texts, enabling the growth of scientific illustration for the first time.

Voyages of discovery

Since a picture is worth a thousand words, the ability to illustrate proved especially important given the strange new reports and objects that would soon flood Europe. This information came from new lands being contacted directly by Europeans. The first source was Asia and sub-Saharan Africa. European contact with these places came about thanks to Portuguese attempts to open a sea route for trade with India in order to cut out the middlemen – predominantly Venetians and Arabs – who controlled the overland and Mediterranean routes. In the early 15th century, the Portuguese prince known as Henry the Navigator (1394–1460) began sending expeditions down the west African coast, establishing direct contact with traders in sub-Saharan Africa. Portuguese sailors pushed on further and further south, eventually rounding the Cape of Good Hope in 1488, and culminating in Vasco da Gama's successful trading voyage to India in 1497–98. The Portuguese established trading outposts all along the route, many of which remained Portuguese possessions until the middle of the 20th century, and eventually extended their regular voyages as far as China, transporting luxury goods like spices, precious stones, gold, and porcelain back to Europe. They also brought back stories of distant lands, strange creatures, and unknown peoples.

This broadening of European horizons did not begin abruptly in the Renaissance. The Middle Ages laid the foundations for Renaissance-era voyages. Indeed, the eastward voyages of the 15th century re-established contacts that had been made in the

13th but cut off in the 14th due to political upheavals in Asia. Medieval travellers, often members of the two new religious orders of the 13th century – Dominicans and Franciscans – embarked on distant religious and ambassadorial missions to an extent we are only now beginning to recognize. They established religious houses across Asia all the way to Peking, as well as in Persia and India, and sent back information to Europe that informed and inspired later mercantile voyages. These medieval travels resulted in a broader sense of the place of Europe within a much larger world to be explored.

While the Portuguese were opening sea routes eastward towards Asia, Christopher Columbus was staring off in the opposite direction. Convinced that the circumference of the earth was about one-third less than the fairly accurate estimates made in antiquity and still widely known in Europe, Columbus imagined that he could reach East Asia faster by sailing westwards. This mistaken impression was in part due to Ptolemy, the 2nd-century geographer and astronomer. Humanists had recently recovered his *Geography*, which included an anomalously small figure for the size of the earth and considerably overestimated the eastward extent of Asia. Financial backers of Columbus were duly sceptical; they recognized that the westward route was the longer way around, and without intermediate places to take on fresh supplies, the crew would starve. (*No one* thought Columbus would 'sail off the edge of the earth', since the sphericity of the Earth had been fully established in Europe for over 1,500 years before Columbus. The notion that people before Columbus thought that the Earth was flat is a 19th-century invention. Medievals would have had a good laugh at the idea!) Hence, when in 1492 Columbus's ships struck land in the Caribbean, he thought he had reached Asia rather than discovered a new continent.

Whether or not Columbus later acknowledged his mistake, others quickly did, and hastened to travel to this New World. News of the new continent spread quickly, aided by the young printing press, and in 1507, a German cartographer gave the new lands a name – America – after the Italian explorer Amerigo Vespucci. Thanks to these maps and Vespucci's accounts of South America published with them, the name stuck. In 1508, King Ferdinando II of Spain created the position of chief navigator for the New World for Vespucci. This new position existed within the Casa de Contratación (House of Trade), a centralized bureau founded in 1503 not only for collecting taxes on goods brought back to Spain, but also for collecting and cataloguing information of all kinds from returning travellers, for training pilots and navigators, and for constantly updating master maps with new information gleaned from every returning ship's captain. The knowledge and practical know-how collected in Seville helped Spain establish the first empire in history upon which 'the sun never set'.

Other nations, not wishing to be left out of the territories and wealth Spain and Portugal were amassing, joined the fray, although trailing the Iberians by a century or more. Thus for a hundred years, virtually all the New World reports and samples that transformed European knowledge of plants, animals, and geography came into Europe through Spain and Portugal. It is hard to imagine the flood of data that poured into Europe from the New World. New plants, new animals, new minerals, new medicines, and reports of new peoples, languages, ideas, observations, and phenomena overwhelmed the Old World's ability to digest them. This was true 'information overload', and it demanded revisions to ideas about the natural world and new methods for organizing knowledge. Traditional systems of classifying plants and animals were exploded by the discovery of new and bizarre creatures. Observations of human habitation virtually everywhere explorers could reach refuted the ancient notion that the world was divided into five climatic regions – two temperate ones and three rendered uninhabitable due to excessive heat or cold. Exploiting the enormous economic potential of the Americas and Asia required fresh scientific and technological skills. Geographical data and the recording of sea routes drove the creation of new mapping techniques, while getting safely and reliably between Europe and the new lands demanded improvements to navigation, shipbuilding, and armaments.

Reforms of Christianity

While voyages around the world exposed Europeans to a diversity of religious perspectives, such perspectives were also diversifying at home. The year 1517 marks the beginning of a deep, often violent, and continuing rupture within Christianity. In that year, the Augustinian priest and theology professor Martin Luther (1483–1546) proposed his famous 'Ninety-Five Theses' in the university town of Wittenberg. These theses, or propositions, were written in the format of topics for Scholastic disputation, and centred on inappropriate and theologically indefensible contemporaneous local practices involving the sale of indulgences. While similar debates over practical and doctrinal issues were common fare in the disputative university culture of the Middle Ages, Luther's protest passed beyond the usual confines of scholarly theological disputation and quickly became a broad-based political and social movement out of Luther's control. Although initially quite mild, Luther's claims became increasingly bold and confrontational, moving from relatively minor issues of local practices into serious doctrinal matters. These claims were quickly disseminated by the printing press, deepened by linkages to local nationalism, and abetted by Germanic rulers who saw separation from Rome as favourable to their political interests. A local protestation thus unexpectedly became Protestantism. Protestantism almost immediately splintered into sparring sects. Catholic-Lutheran controversies were soon joined by Lutheran-Calvinist ones, then by intra-Calvinist ones, and so on. The so-called 'Wars of Religion' – often motivated more by political and dynastic manoeuvres than by doctrinal issues – convulsed Europe, particularly Germany, France, and England, for the next century and half.

Luther himself was no humanist, although some of his notions, such as an emphasis on a literal reading of the Bible as opposed to the allegorical readings favoured by Catholics, bear resemblances to humanist emphases on texts. But these resemblances are outweighed by his suspicion of Classical ('pagan') literature and ideas and his desire to expunge books from the Bible (such as the Letter of James) that disagreed with his personal notions. The much more learned Philipp Melanchthon (1497–1560), however, was quite a different story. Melanchthon's very name testifies to his humanism, translated into Classical Greek from the original barbarous German Schwartzerd ('black earth'). His great uncle, Johannes Reuchlin, who suggested this 'self-classicization', was the most prominent humanist in Germany. In the wake of the Lutheran rejection of university Scholasticism, Melanchthon (who as a humanist also disliked Scholasticism) renovated university curricula and pedagogy in German universities – in particular, Luther's own University of Wittenberg – as they converted from Catholic to Lutheran. The new curricula he devised earned him the title *Praeceptor Germaniae* ('Teacher of Germany'). His approach was not to banish Aristotle, but rather – in true humanist fashion – to banish medieval 'accretions' to Aristotle and to use better editions of the Greek philosopher. New Protestant universities found themselves in the enviable position of having to start afresh, that is, with a reduced burden of established methods, and were thus able to incorporate new subjects and approaches that had not found a place in older institutions.

Within Catholicism, reform movements were also underway. In the 15th century, church councils addressed some issues, although not very successfully. More dramatic was the Council of Trent (1545–63), an Ecumenical Council convened to respond to Protestantism by addressing corruption, clarifying doctrines, standardizing practices, and centralizing disciplinary oversight. The Council of Trent, the most important post-medieval church council until Vatican II (1962–5), launched the Catholic Reform, or 'Counter-Reformation'. Its measures included improved education for priests, a reform many humanists had been advocating, but also increased oversight of orthodoxy including in published works. Tridentine reforms were taken up most avidly by a newly organized society of priests, the Society of Jesus, or Jesuits. Organized by St Ignatius Loyola and given papal authorization in 1540, the Jesuits devoted themselves especially to education and scholarship, and made significant contributions specifically to science, mathematics, and technology.

The broader impact of the Jesuits, besides preaching for a return of Protestants to Catholicism, lay in the hundreds of schools and colleges they established within the first years of their existence. Jesuit pedagogy rested upon an innovative style of teaching and curriculum, one that preserved the importance of Aristotelian methods, but paired that with new emphasis on mathematics (by 1700, more than half of all the professorships of mathematics in Europe were held by Jesuits) and the sciences. Jesuit schools were often the first to teach some of the new scientific ideas of the Scientific Revolution, and educated many of the thinkers responsible for them. Jesuits spread out across the globe along the newly opened trade routes, establishing a high-profile presence (and schools, of course) in China, India, and the Americas, and the first global correspondence network. This network channelled everything from biological specimens and astronomical observations to cultural artefacts and extensive reports of native knowledge and customs back to Rome. The Jesuit attitude in studies of science and mathematics expresses their motto 'to find God in all things'. While Jesuits emphasized this incentive, it was not unique to them – it undergirded virtually the entire Scientific Revolution.

The new world of the 1500s

Europeans of the 16th century inhabited a new and rapidly changing world. As in our own fast-paced days, many saw this situation as a source of anxiety, while others saw a world of opportunities and possibilities. The horizons of Europe had been expanded in every sense. Europeans had rediscovered their own past, encountered a wider physical and human world, and created new approaches and fresh interpretations of older ideas. Indeed, the best image for their world would be that of a tumultuous and richly stocked market place. A cacophony of voices promoted a diversity of ideas, goods, and possibilities. Throngs jostled elbows to test, purchase, reject, praise, criticize, or just touch the varied merchandise. Almost everything was up for grabs. Whether we conclude the 'Scientific Revolution' to be something entirely new, or a revival of the intellectual ferment of the late Middle Ages after the interruption of the baleful 14th century, there can be no doubt that the learned inhabitants of the 16th and 17th centuries saw their time as one of change and novelty. These were exciting times; times of new worlds indeed.

Chapter 2
The connected world

When early modern thinkers looked out on the world, they saw a *cosmos* in the true Greek sense of that word, that is, a well-ordered and arranged whole. They saw the various components of the physical universe tightly interwoven with one another, and joined intimately to human beings and to God. Their world was woven together in a complex web of connections and interdependencies, its every corner filled with purpose and rich with meaning. Thus, for them, studying the world meant not only uncovering and cataloguing facts about its contents, but also revealing its hidden design and silent messages. This perspective contrasts with that of modern scientists, whose increasing specialization reduces their focus to narrow topics of study and objects in isolation, whose methods emphasize dissecting rather than synthesizing approaches, and whose chosen outlooks actively discourage questions of meaning and purpose. Modern approaches have succeeded in revealing vast amounts of knowledge about the physical world, but have also produced a disjointed, fragmented world that can leave human beings feeling alienated and orphaned from the universe. Virtually all early modern natural philosophers operated with a wider, more all-embracing vision of the world, and their motives, questions, and practices flowed from that vision. We have to understand their worldview if we are to understand their motivations and methods in investigating that world.

The concept of a tightly connected and purposeful world derives from many sources, but above all from the two inescapable giants of antiquity, Plato and Aristotle, and from Christian theology. From Platonic sources, particularly the thinkers called Late Platonists or Neoplatonists – philosophers actively developing Plato's ideas in Hellenized Egypt during the first centuries of the Christian Era – comes the idea of a *scala naturae*, or ladder of nature. According to this conception, everything in the world has a special place in a continuous hierarchy. At the very top is the One – the utterly transcendent, eternal God, from whom everything else derives existence. The One emanates creative power that brings everything else into existence. The further this power radiates from its Source, the lower and more unlike the One are the things it creates. At the bottom lies inert, lifeless matter. The rungs in between, in ascending order, are filled with vegetable and animal life, then human beings, and then spiritual beings such as *daimons* and lesser gods. The goal of some Neoplatonists was to climb the ladder as it were, to became more spiritual and less material, to free the human soul – our most noble part – from the blindness caused by its descent into matter, and to rise through the levels of spiritual beings in journey towards the One. This late antique conception both influenced and was influenced by Christian doctrines, and could be readily adapted to orthodox Christian beliefs by replacing the pagan *daimons* and lesser gods with orders of angels, and the One with the Christian God, as was suggested by the 5th-century Christian Neoplatonist pseudo-Dionysius the Areopagite. Thanks to such Christianization, the idea of the *scala naturae* remained well known throughout the Latin Middle Ages, even if the ancient Platonic texts upon which it was based were lost for centuries.

These Platonic texts were among those rediscovered by humanists in the Renaissance and translated by Marsilio Ficino. Ficino also acquired, translated, and published a set of texts attached to the name Hermes Trismegestus, meaning Hermes 'the Thrice-Great', a supposed ancient Egyptian sage contemporary with Moses.

What Ficino obtained was a small selection out of a huge mass of diverse *Hermetica* (writings attributed to Hermes) dating from about the 3rd century BC to the 7th AD. Although initially believed to be much older, Ficino's *Hermetica* probably dates from the 2nd and 3rd centuries AD. Its importance lies in its Neoplatonic character that emphasizes the power of human beings, their place in the connected world of the *scala*, and their ability to ascend it. Many Renaissance readers found what they thought to be foreshadowings of Christianity in the *Hermetica*, and thus Hermes Trismegistus took on the status of a pagan prophet, and accordingly he can be found depicted among the prophets in the cathedral of Siena.

The *scala* envisions of a world in which every creature has a place, and each creature is linked to those immediately above and below it, such that there is a gradual and continuous rise from the lowest level to the highest, without gaps, along what has been called 'the Great Chain of Being'. A related concept – present in the *Timaeus*, Plato's account of the origin of the universe, and the only work of Plato known to the Latin Middle Ages – is that of the *macrocosm* and *microcosm*. These two Greek words mean, respectively, the 'large ordered world' and the 'little ordered world'. The macrocosm is the body of the universe, that is, the astronomical world of stars and planets, while the microcosm is the body of the human being. The essential idea is that these two worlds are constructed on analogous principles, and so bear a close relationship to each other. A late contribution to the *Hermetica*, an 8th-century Arabic work called the *Emerald Tablet*, concisely summarizes this view in a terse motto well known in early modern Europe: 'as above, so below'. For Plato, the linkage of man's microcosm with the planetary macrocosm had a practical moral meaning – we should look to the orderly, rational workings of the heavens as a guide for governing ourselves in an orderly, rational way. For early modern Europeans, the microcosm–macrocosm linkage had, above all, a medical meaning – it undergirded medical astrology. The various planets have particular effects upon particular human organs, whereby they can influence the bodily functions (see Chapter 5).

A second major contributor to the view of an interconnected and purposeful world comes from Aristotelian ideas about how to gain knowledge. According to Aristotle, proper knowledge of a thing is 'causal knowledge'. That term requires explanation. Aristotle argued that knowing a thing requires identifying its four 'causes', or reasons for existing. The first of these, the *efficient cause*, describes what or who made the thing. The *material cause* describes what the thing is made of. The *formal cause* tells what physical characteristics make the thing what it is, in other words, an inventory of its qualities. The most important cause for Aristotelians, and the most difficult one for moderns to get their minds around, is the *final cause*. The final cause tells what the thing is for, that is, what its goal in existing is, and for Aristotle, everything has a goal or purpose. These 'causes' can be illustrated using a statue of Achilles. The statue's efficient cause is the sculptor, its material cause is marble, its formal cause is the beautiful body of Achilles, and its final cause is to celebrate the memory of Achilles. There can be more than one of each of the causes (for example, the statue might also have the final cause of being decorative, or perhaps, in some Attic house, to act as a coat rack).

The crucial point is that Aristotelian forms of knowledge, particularly in regard to the efficient and final causes, acted to define objects *in the context of their relationship to other objects*. Coming to know a thing meant being able to position it within a network of relationships with other things, particularly the things that bring it into being and that make use of it. In the Christian context of Europe, the final cause harmonized well with the idea of divine design and providence. Final causes in nature were part of God's plan for creation, implanted and encoded within created things by the First Efficient Cause.

Writers of the early modern period expressed their understanding of a connected world in many different ways. The English natural philosopher Robert Boyle (1627–91), renowned for his work in chemistry (chemistry students still have to learn Boyle's Law that the volume of a gas is inversely proportional to the pressure exerted upon it), wrote that the world is like 'a well contrived Romance'. Here, Boyle alludes to the massive French novels of his day (of which he was very fond). These romances often run to more than two thousand pages in length, and feature a memory-taxing myriad of characters whose complex storylines constantly converge and diverge in surprising ways, full of revelations about who is secretly in love with whom and who is really whose long-lost brother, child, or what-not. For Boyle, the Creator is the ultimate romance writer, and scientific investigators are the readers trying to figure out all the relationships and crisscrossing storylines in the world He wrote.

The Jesuit polymath Athanasius Kircher (1601/2–80), who maintained a museum of wonders in Rome and was a centre of Jesuit correspondence about natural philosophy, portrayed the connected world in an elegant Baroque frontispiece to his encyclopaedic work on magnetism (Figure 1).

The image shows a series of circular seals, each bearing the name of one branch of knowledge: physics, poetry, astronomy, medicine, music, optics, geography, and so on, with theology at the top. A single chain connects the seals together, expressing the inherent unity of all branches of knowledge. For early moderns, there were no strict barriers that kept sciences, humanities, and theology insulated from one another – they formed interlocking ways of exploring and understanding the world. In Kircher's image, these branches of knowledge stand chained to three larger seals representing the three chief parts of the natural world: the siderial world (everything farther away than the Moon), the sublunar world (the Earth and its atmosphere), and the microcosm (human beings). These three parts of the world are likewise

1. Engraved title page to Athanasius Kircher, *Magnes sive de magnetica arte* (Rome, 1641) expressing the interconnectedness of the branches of knowledge and of God, humanity, and nature

chained together indicating the inescapable interdependence that exists between them. At the centre of the entire image, in direct contact with each one of the three worlds equally, stands the *mundus archetypus* – the archetypal world, that is, the mind of God that not only created everything, but also contains within itself the models or archetypes of everything possible in the universe. Kircher completes his image with the Latin motto: 'Everything rests placidly, connected by hidden knots.'

This sense of connectedness both between disciplines and between various facets of the universe characterizes *natural philosophy* – the discipline practised by early modern students of the natural world. Natural philosophy is closely related to what we familiarly call *science* today, but is broader in scope and intent. The natural philosopher of the Middle Ages or of the Scientific Revolution studied the natural world – as modern scientists do – but did so within a wider vision that included theology and metaphysics. The three components of God, man, and nature were never insulated from one another. Natural philosophical outlooks gradually gave way to more specialized and narrow 'scientific' ones only during the 19th century (the age in which the word 'scientist' was first coined). The work and motivations of early modern natural philosophers cannot be properly understood or appreciated without keeping the distinct character of natural philosophy in mind. Their questions and goals were not necessarily our questions and goals, even when the very same natural objects were being studied. Hence, the history of science cannot be written by pulling scientific 'firsts' out of their historical context, but only by seeing with eyes and minds of our historical characters.

Natural 'magic'

The 'cosmic' perspective was widely shared in the 16th and 17th centuries, and it undergirded a variety of practices and projects, even if different thinkers considered the interconnections in the world to be of varying degrees of importance to their work. The facet of natural philosophy most closely tied to this vision of the world was *magia naturalis*. It is misleading to translate this Latin term directly into English as 'natural magic'. The word 'magic' naturally makes modern readers think of costumed men pulling rabbits out of hats, or of wizened black-robed characters in pointy hats mumbling over cauldrons, or, rather more benignly, of Harry Potter and Hogwarts. The *magia naturalis* of the early modern period was, however, something very different; it forms an important part of the history of science.

Magia is perhaps best translated for moderns as 'mastery'. The goal of the practitioner of *magia*, called a *magus*, is to learn and to control the connections embedded in the world in order to manipulate them for practical ends. Look again at Kircher's frontispiece. In the upper left-hand corner, *magia naturalis* is listed among the branches of knowledge, between arithmetic and medicine. Kircher symbolizes it with the turning of a sunflower to follow the Sun across the sky throughout the day. (Several plants display this behaviour, known as *heliotropism*.) Why does the sunflower always turn towards the Sun while most plants do not? Clearly, there must be some special link between Sun and sunflower. The ability of the sunflower to follow the Sun provided a prime example of the hidden connections and forces in the world that the magus endeavoured to identify and control.

Medieval Aristotelians divided properties of a thing into two groups. The first were *manifest qualities* – qualities that anybody endowed with sense organs could detect. Hot, cold, wet, and dry were the primary qualities. Other qualities included things like smooth, rough, yellow, white, bitter, salty, sonorous, fragrant, and so forth – all things that activated the senses. After all, Aristotelianism was fundamentally a common-sense way of engaging with the world. Aristotelians used these manifest qualities to explain the action of one thing upon another: cooling drinks lower a fever because cold counteracts hot, for example. But some objects acted in weird ways that manifest qualities could not explain. These objects were held to have *hidden qualities* (*qualitates occultae*, often misleadingly translated as 'occult qualities') that we cannot detect with our senses. These qualities often acted in highly specific ways, suggesting a special, invisible connection between specific things and the objects they acted upon. Medieval natural philosophers compiled lists of such phenomena. One classic example is the magnet. We can sense nothing about the lodestone (a naturally magnetic mineral) that could possibly explain its mysterious ability to attract iron specifically. The same is true of the apparent attraction between the Sun and the sunflower, the turning of a compass needle towards the pole star, the sleep-inducing effect of opium, the Moon's effect on the tides, and many other things. *Magia naturalis* was the endeavour to seek out these hidden qualities of things and their effects, and to make use of them.

How did one go about finding these connections, these 'hidden knots', in nature? One way was to observe the world closely. Everyone can agree that careful observation is a crucial starting point for scientific investigation; the pursuit of *magia naturalis* promoted such observation. A method of equal importance lay in mining the records of earlier observers of nature – accounts and observations, ranging from the commonplace to the bizarre, recorded in various texts from contemporaneous times back to the ancient world. Much *magia* was therefore based on a careful reading of texts in humanist fashion, building up complex networks by compiling claims from earlier writers. Given the immense variety of nature, the task of the aspiring *magus* is mind-bogglingly immense – no less than cataloguing the properties of everything. Could there be a shortcut? Some natural philosophers believed that nature contained clues to guide the magus, perhaps as hints implanted there by a merciful God who wants us to understand His creation and benefit from it. The *doctrine of signatures* claims that some natural objects are 'signed' with indications of their hidden qualities. Often, this means that two connected objects look somehow similar, or have some analogous characteristics; for example, the sunflower not only follows the Sun, its blossom actually *resembles* the Sun in colour and shape. Various parts of plants resemble various parts of the human body; a walnut nestled in its shell looks remarkably like a brain inside the skull. Is this a sign that walnuts would provide good medicines for the brain? The practitioner of *magia* would have to try these things out to be sure, but observation coupled with the idea of signatures provided a useful point of departure for investigating, explaining, and using the natural world.

The doctrine of signatures represents but one facet of a broader mode of analogical thinking ubiquitous in the early modern period. While moderns would tend to see such similarities as mere coincidence or accident, or as 'poetic' rather than physical, many early moderns saw things quite differently – they *expected* analogical links between different parts of the world, and the discovery of an analogy or symmetry in nature signified for them a real connection between things. Rather than being the product of human imagination, every analogy between two objects in the natural world marked out another line in the blueprint of creation, a visible sign of a hidden connection divinely implanted in the universe. Thus, arguments from analogy carried special strength and evidentiary power beyond what we are accustomed to give them today. The sureness of this linkage was founded upon an unshakable faith in a cosmos that was not random or fortuitous, but rather one that was suffused with meaning and purpose, guided in various ways by divine wisdom and providence for the benefit of human beings. This certainty, and the attendant use of analogical reasoning, was not the exclusive property of those interested in *magia naturalis*, but of virtually *every* serious thinker of the period.

Using direct observation, analogy, textual authorities, and signatures, early modern thinkers compiled huge aggregates of things they considered to be linked. For example, what else might relate to the Sun–sunflower connection? The Sun is the source of warmth and life in the macrocosm, its counterpart in the microcosm must be the heart. (Have yet another look at Kircher's frontispiece – there is a tiny Sun in the place of the heart in the human figure representing the microcosm.) The Sun is the most noble of the heavenly bodies, brilliant and yellow, and thus it bears a similitude to gold in the mineral realm, and further afield to all yellow or golden things. In the animal realm, the Sun causes the rooster to crow, indicating a special link between the two. The lion, with its tawny colour, royal status, and head that resembles the Sun (its mane frames its head like solar rays), also seems linked to the Sun. Likewise, the bravery of the lion corresponds in turn with the heart. Sun, sunflower, heart, gold, yellow, rooster, and lion all bear links of commonality and thus real but hidden connections. For the advocates of *magia naturalis*, these analogical links translate into operative links that can be put to use. The most down-to-earth application would involve using gold or sunflowers to make a medicine for the heart – but things could get much more dramatic, as we shall see.

Opinions varied as to what actually linked objects bound up in these webs of correspondence, but they were usually considered to function by means of 'sympathy', which literally means 'suffering together or receiving action together'. Think of two well-tuned lutes on opposite sides of a room, pluck a string on one of them, and the corresponding string on the other will immediately start to vibrate and hum on its own, echoing the note plucked on the first lute. Today, we still call this phenomenon *sympathetic* vibration. For early modern thinkers, this phenomenon exemplified the operation of unseen links acting at a distance between things that were 'in tune' with one another. Some argued that a medium was necessary to transmit the action between spatially separated objects; Aristotle had argued that one thing could not act on another thing a distance away without an intervening medium to carry the effects. In the case of lute strings, for example, we know that the intervening air carries the vibrations between the two instruments. For other sympathetic actions, this medium might be the so-called *spiritus mundi*, or spirit of the world – a universal, all-penetrating incorporeal or quasi-corporeal substance, capable of keeping even distant objects in virtual contact with one another by transmitting influences from one to the other. This 'spirit' was not some sentient supernatural entity; rather, it is the macrocosmic equivalent of the microcosmic animal spirits, the subtle substance in our bodies that transmits the command 'move!' through the nerves to our feet when our intellect realizes that a two-ton truck is speeding towards us. The spirit of the world likewise carries 'signals' from the Sun to the sunflower or from the Moon to the waters of the sea. Once again, the microcosm and the macrocosm are reflections of one another; both contain spirits that transmit signals. Incidentally, this analogous nature should also mean that the macrocosm itself has a soul of some sort – a point Plato asserts in the *Timaeus* and is especially difficult for moderns to understand – the next chapter returns to this point.

Practical 'mastery' from the kitchen to the study

The theory of natural magic in regard to a connected world is impressive, even elegant and beautiful, yet the key feature of *magia naturalis* is practical application. The practical parts of early modern *magia* range from the banal to the sublime, the former often having little to do with any theoretical foundations. The book *Magia naturalis* of Giambattista della Porta (1535–1615) provides a good example. Della Porta is renowned for establishing in Naples the earliest scientific society – the Academy of Secrets – and for being a member of the Accademia dei Lincei, the early 17th-century scientific society that counted Galileo as a member. The first chapter of Della Porta's book recapitulates the principles of an interconnected world, noting how magic 'is the survey of the whole course of nature' and 'the practical part of natural philosophy'. He advises his reader to 'be prodigal in seeking things out; and while he is busy and careful in seeking, he must be patient also ... neither must he spare any pains: for the secrets of nature are not revealed to lazy and idle persons'. The practical secrets of nature that the rest of della Porta's book reveals do include observations about magnetism and optics, but the majority of the book is a miscellany of recipes for everything from making artificial gems and fireworks, to animal and plant breeding, to household hints about making perfumes, roasting meat, and preserving fruit, none of which draws upon any theoretical conception of the world. Della Porta's book fits instead with a tradition of 'books of secrets' that became increasingly popular throughout the 16th and 17th centuries, some of which were reprinted even into the 19th. Many such books begin with an exposition of grand and lofty notions about the cosmos, but consist principally of recipes for household management or cottage industries, and contain little or nothing about the nature of the world.

At the sublime end of the scale stands Marsilio Ficino (1433–99), whose practical application of the connectedness of the world was expressed in ways of living and in rituals. Ficino often complained of his melancholy temperament; perhaps he suffered from what we now label as depression. The established medicine of the day held that a preponderance of black bile – one of the four 'humours' of the body that must remain in balance to provide health – produces depression. Indeed, the Greek term for black bile – *melaina cholē* – is the origin for our word *melancholy*. (In the same way, personalities that are still called sanguine, choleric, and phlegmatic arise from the preponderance of one of the other three bodily humours: blood, yellow bile, or phlegm, respectively; see Chapter 5.) Ficino explored the connection between the scholarly life and melancholy, and proposed lifestyle changes for his fellow intellectuals to help them address the problem. He formulated a diet and medicinal supplements to prevent the formation of excess black bile in the body, and his 'On Obtaining Life from the Heavens' proposes using celestial influences to counteract this occupational hazard of scholars.

Physicians considered black bile to have the manifest qualities of cold and dry. The planet Saturn shares these qualities, and thus the two bear a sympathetic connection. Therefore, anything in the web of correspondences with black bile and Saturn was to be avoided. The opposing qualities of the Sun (hot–dry) and Jupiter (hot–wet) counteract the cold–dry of black bile, and so by analogical extension anything in the web of correspondences with the Sun and Jupiter could help counteract scholarly melancholy. (Our word 'jovial' literally means 'relating to Jupiter', an indication preserved in our language of how thoroughly entrenched and accepted this reasoning really was.) Thus, in order to make use of sympathetic links to the Sun, the Florentine humanist suggested wearing yellow and golden clothes, decorating one's chamber with heliotropic flowers, getting lots of sunlight, wearing gold and rubies, eating 'solar' foods and spices (like saffron and cinnamon), hearing and singing harmonious and stately music, burning myrrh and frankincense, and drinking wine in moderation. For some readers, however, he did tread a little too far when he also suggested – following the lead of the ancient Neoplationists Plotinus and Iamblichus, whose works he translated from Greek – making images that could attract and capture planetary powers, a rather questionable thing for an ordained Roman Catholic priest to be doing. Indeed, Ficino can be read as crossing the line at this point from *natural* magic into *spiritual* magic, although he might well have disputed that interpretation. The former used the hidden sympathies in nature, while the latter elicited the help of spiritual beings – the *daimons* and gods of pagan Greek philosophy, or the demons and angels of Christian theology. The former *magia* was unobjectionable, the latter (reasonably enough) drew the condemnation of theologians. Questions were raised about Ficino's orthodoxy, but apparently no actions were taken, since such rituals could be interpreted as entirely physical and medicinal, and thus entirely acceptable. Over a century later, for example, the Dominican friar Tommaso Campanella and Pope Urban VIII used a ritual of lights, colours, smells, and sounds, not unlike Ficino's prescriptions, to counteract any possible ill effects from the temporary loss of healthful solar influences during a solar eclipse that had been predicted to bring about the pontiff's death. The Pope survived. Yet while this *magia* was natural in intended operation, some onlookers did view such applications as suspect.

At the present time, applications of *magia naturalis* and the whole idea of an interconnected world of sympathies and analogies are sometimes dismissed as irrational or superstitious. But this harsh judgement is faulty. It results from a certain smug arrogance and a failure to exercise historical understanding. What our predecessors did was to observe various mysterious and apparently similar phenomena in nature and to extrapolate thence into a more universal statement – a law of nature – about connections and the transmission of influences in the world. This extrapolation led to one tenet that they held that we do not; namely, that similar or analogous objects silently exert influence upon one another. Once that assumption is made, then the rest of the system builds upon it rationally. They were trying to understand the world; they were trying to make sense of things and to make use of the powers of nature. They moved inductively from observed or reported instances to a general principle and then deductively to its consequences and applications. We might choose to say, informed as we are by more recent studies, that the action between Sun and sunflower, or Moon and sea, or magnet and iron, can be better explained by something other than hidden knots of sympathy. But that does not permit us to say that their methods or conclusions were irrational, or that the beliefs and practices that came from them were 'superstitious'. If that leap were allowed, then every scientific theory that comes ultimately to be rejected in the course of the development of our understanding of the world – no doubt including some things that we today believe to be true explanations of phenomena – would have to be judged irrational and superstitious as well, rather than simply *mistaken* notions that were arrived at rationally given the ideas, perspectives, and information available at the time.

Religious motivations for scientific investigation

Magia naturalis is only the strongest expression of widely held ideas of a connected world, of the macrocosm and microcosm, and of the power of similitude. The same kinds of connections and thinking were often implicit in the work of natural philosophers who never gave natural magic a second thought. Every thinker of the period, for example, was confident of the intimate connections among human beings, God, and the natural world, and consequently of the interconnections between theological and scientific truths. This feature brings up the complex topic of science and theology/religion. In order to understand early modern natural philosophy, it is necessary to break free of several common modern assumptions and prejudices. First, virtually everyone in Europe, certainly every scientific thinker mentioned in this book, was a believing and practising Christian. The notion that scientific study, modern or otherwise, requires an atheistic – or what is euphemistically called a 'sceptical' – viewpoint is a 20th-century myth proposed by those who wish science itself to be a religion (usually with themselves as its priestly hierarchy). Second, for early moderns, the doctrines of Christianity were not opinions or personal choices. They had the status of natural or historical facts. Dissension obviously existed between different denominations over the more advanced points of theology or ritual practice, just as scientists today argue over finer points without calling into question the reality of gravity, the existence of atoms, or the validity of the scientific enterprise. Never was theology demoted to the status of 'personal belief'; it constituted, like science today, both a body of agreed-upon facts and a continuing search for truths about existence. As a result, theological tenets were considered part of the data set with which early modern natural philosophers worked. Thus theological ideas played a major part in scientific study and speculation – not as external 'influences', but rather as serious and integral parts of the world the natural philosopher was studying.

Many people today acquiesce in the widespread myth, devised in the late 19th century, of an epic battle between 'scientists' and 'religionists'. Despite the unfortunate fact that some members of both parties perpetuate the myth by their actions today, this 'conflict' model has been rejected by every modern historian of science; it does not portray the historical situation. During the 16th and 17th centuries and during the Middle Ages, there was not a camp of 'scientists' struggling to break free of the repression of 'religionists'; such separate camps simply did not exist as such. Popular tales of repression and conflict are at best oversimplified or exaggerated, and at worst folkloristic fabrications (see Chapter 3 on Galileo). Rather, the investigators of nature were themselves religious people, and many ecclesiastics were themselves investigators of nature. The connection between theological and scientific study rested in part upon the idea of the Two Books. Enunciated by St Augustine and other early Christian writers, the concept states that God reveals Himself to human beings in two different ways – by inspiring the sacred writers to pen the Book of Scripture, and by creating the world, the Book of Nature. The world around us, no less than the Bible, is a divine message intended to be read; the perceptive reader can learn much about the Creator by studying the creation. This idea, deeply ingrained in orthodox Christianity, means that the study of the world can itself be a religious act. Robert Boyle, for example, considered his scientific inquiries to be a type of religious devotion (and thus particularly appropriate to do on Sundays) that heightens the natural philosopher's knowledge and awareness of God through the contemplation of His creation. He described the natural philosopher as a 'priest of nature' whose duty it was to expound and interpret the messages written in the Book of Nature, and to gather together and give voice to all creation's silent praise of its Creator.

In sum, early moderns saw – in various ways – a cosmically interconnected world, where everything, human beings and God and all branches of knowledge, were inextricably linked parts of a whole. In some respects, the recent development of ecology and environmental sciences might be seen as restoring some lines of the unseen networks of interdependence early modern natural philosophers envisioned in their own world. However that may be, early modern thinkers, like their medieval forebears, looked out on a world of connections and a world full of purpose and meaning as well as of mystery, wonder, and promise.

Chapter 3
The superlunar world

Until the modern age, the heavens were quite literally half of people's daily world. The sky and its movements were inescapable. It is ironic and tragic that while modern science now gives us better explanations of the workings of the celestial world than ever before, modern technology means that most people can no longer see its nightly movements with their own eyes, feel its presence, and marvel at its beauty. It now requires an unobscured view far from the pollution of light and industry to witness the impact of the night sky as our ancestors did. Long before the invention of writing, ancient peoples knew the movements of the heavens. Figuring out how to explain these movements, however, occupied acute minds down to the 18th century. The gradual uncovering of the hidden structures of the heavens represents a key narrative of the Scientific Revolution. The best-known names of the era – Copernicus, Kepler, Galileo, Newton – are principal players in this story. Indeed, developments in astronomy stood for a long time as *the* narrative for the period, providing much of the foundation for giving it the title of 'revolution'.

For the intellectual of 1500, the universe was divided into two realms: the *sublunar world* of the Earth and everything up to the Moon, and the *superlunar world* of the Moon and everything beyond. This division had been drawn by Aristotle, based on the common observation of the dichotomy between the unchanging heavens and the ever-changing Earth. In the sublunar world, the four elements of earth, water, air, and fire constantly combine, dissociate, and recombine; new things appear and old things vanish. The superlunar world was quite a different matter; it was the realm of the unchanging. For centuries before Aristotle, stargazers had watched planets and stars follow their courses with perfect regularity. This absence of change suggested to Aristotle that the superlunar world was composed of a single homogeneous substance, a fifth element he called *aither* (later writers called it quintessence), which could neither change nor decompose because it was pure and elemental.

Observational background

The Greeks initiated the long endeavour to *explain* celestial motions physically and mathematically. These motions are more complex and more orderly than most people today recognize. Everyone is familiar with the daily motion of rising and setting. Everything – Sun, Moon, planets, stars – rises and sets once a day, moving east to west across the sky. Other celestial motions demand more patient observation. The stars, called the 'fixed stars' because they do not move relative to one another, take a little less than 24 hours to come back to the same position in the sky. That means that each star rises a little earlier (about four minutes) each night; therefore, if you look at the sky every night at the same time, you will see the constellations moving slowly from night to night in great arcs around – if you are in the Northern Hemisphere – the one star that never moves, Polaris, the pole star, found at the end of the Little Dipper (Ursa Minor). The stars take one year to return to the same place in the sky at the same time of night. The impression is that of a great shell studded with stars, turning around the Earth once every 23 hours and 56 minutes.

The Sun moves a little more slowly, taking a full 24 hours for one revolution, meaning that from day to day it changes its position relative to the stars, moving slowly *west to east relative to the backdrop of the stars*, taking one year before it lines up with the same stars again. The Moon makes a similar motion, but much more noticeably. It rises about 50 minutes *later* each night, so if you look for it at the same time on consecutive nights, you will find it further to the east every night. (Go ahead and try it!) After 29 days, the Moon is back where it started. The planets do the same thing, but with a weird twist that screams out for explanation. Most of the time they act like Sun and Moon, moving slowly west to east against the backdrop of stars. But at intervals, they slow down, stop, turn around, and move in the opposite direction, going now east to west. This is called *retrograde motion*. After a while, they stop again, turn around, and resume their usual motion.

The ancient Greeks gave the name 'planet' (meaning 'wanderer') to all seven heavenly bodies that appeared to move against the fixed background of stars: the Sun, the Moon, Mercury, Venus, Mars, Jupiter, and Saturn. But the planets don't wander far; their movements are restricted to a narrow band in the heavens called the zodiac. The zodiac is divided into twelve sections of equal length, each containing a single constellation or 'sign': Aries, Taurus, Gemini, and so on. Thus, as the planets make their individual motions against the backdrop of the stars, they appear to move through the zodiac from one constellation and from one sign to the next. A person's 'sign' is whatever zodiacal sign the Sun was 'in' on the day the person was born. But more on astrology in a little while.

Historical background

Plato was convinced that the heavens moved according to harmonious mathematical laws. He was inspired by the ideas of the Pythagoreans, a secretive religious community, who taught that mathematics – number, geometrical shape, ratio, and harmony – was the proper foundation of both the universe and the well-governed life. For Plato and those he inspired down to the modern age, the Creator is a geometer. But the irregular motions of the planets seemed discordant with the idea of a well-regulated mathematical world. Plato therefore argued that their motion only *appears* irregular, and that there exists a divine regularity hidden from our eyes. Because he considered the circle to be the most perfect and regular shape, and motion in a circle to be without beginning or end and thus eternal, he challenged his students to explain the apparent motions of the planets using combinations of *uniform circular motions*. That challenge inspired astronomers for over two thousand years.

Plato's student Eudoxus proposed a universe built up of concentric spheres, like layers of an onion, with the Earth at the centre. Each sphere rotated uniformly, but each planet received the combined motion of several spheres, that added up (approximately) to the observed motion. Eudoxus' system was a *mathematical* model. He did not worry about how the heavens worked physically, or whether there really were spheres up there. The point was to account for observations mathematically. Aristotle, however, wanted a *physical* model. He made Eudoxus' spheres real, solid objects that literally carried the planets around, and accounted for how motion could be transferred from one sphere to the next, like gears of a celestial clockwork. His achievement was to construct an astronomy and physics that worked together harmoniously (Figure 2).

The problem with the concentric spheres model was that it failed to explain observations accurately. For example, the planets change in brightness, as if they were sometimes closer and sometimes farther away, and the seasons are not of equal length. Neither is it possible if the planets are carried by spheres centred on the Earth (Figure 3).

Later astronomers addressed these problems, culminating in the system of Claudius Ptolemy (c. AD 90–c. 168). To solve the problem of unequal seasons, Ptolemy used an *eccentric*; that is,

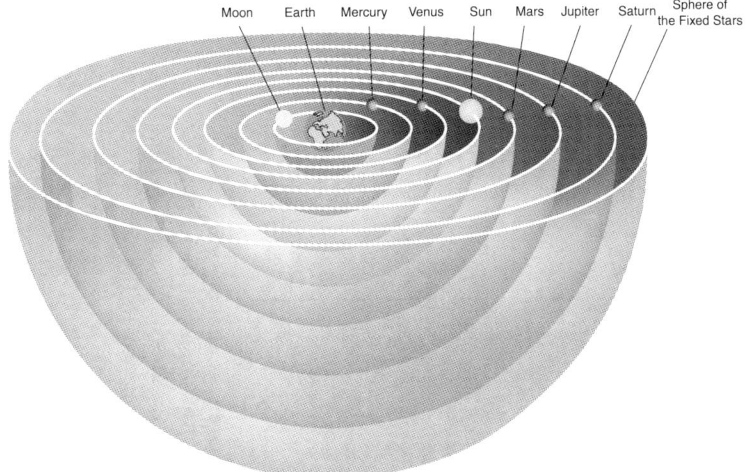

2. A simplified version of Aristotle's concentric spheres model in cross-section

he moved the Earth off-centre. In his system, each sphere has its own centre, none of them coincident with the Earth.

To account for planetary positions better and to solve the problem of their changing brightness, Ptolemy used *epicycles* (Figure 4). Each planet moves in a small circular path centred on, and carried around by, a larger sphere (the deferent) around the Earth. The motions of epicycle and deferent combine to give the planet a looping path that explains observed motions extremely well, and in which the planet is sometimes closer to the Earth, hence brighter.

Ptolemy's system gave good predictions of planetary positions but satisfied the mathematically inclined more than the physically inclined. Aristotle's physics held that heavy bodies fall towards the centre of the universe, which is why a spherical Earth occupies that space and why heavy objects fall. But Ptolemy's Earth is

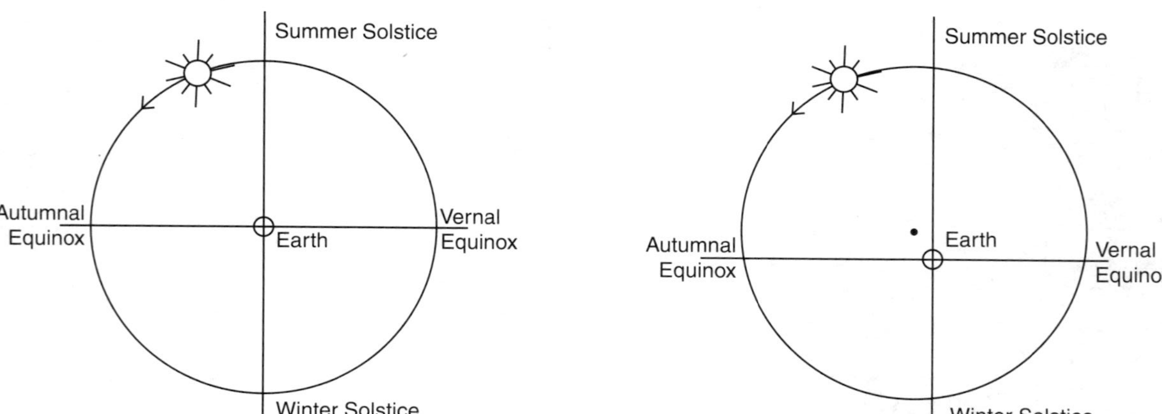

3. (left) If the Earth were at the centre of the Sun's sphere, the Sun's apparent annual motion would divide into four equal arcs, making the seasons have equal lengths. But summer is in fact longer than winter; (right) Ptolemy's off-centre Earth divides the Sun's path into four arcs of unequal length, corresponding correctly with the unequal seasons. This arrangement also explains why the Sun appears to move more slowly in the summer: because it is farther away then

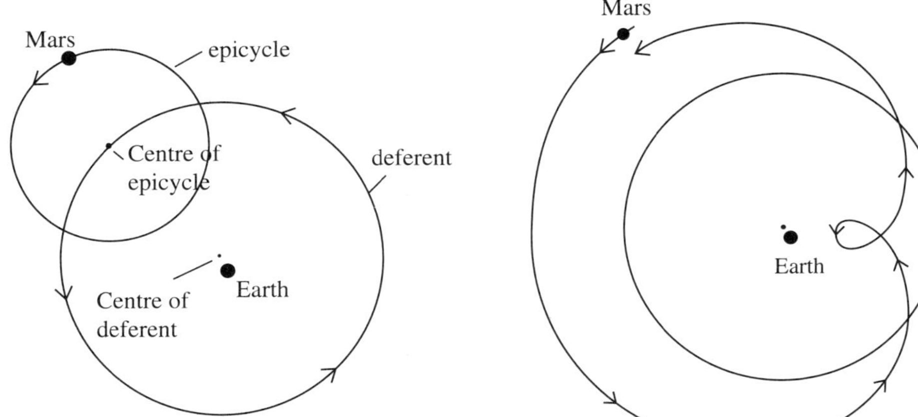

4. (left) A Ptolemaic epicycle and deferent for a planet. The planet moves counterclockwise (looking down at the Earth's north pole) on the epicycle, as the epicycle is carried around on the deferent, also counterclockwise; (right) The apparent motion of the planet resulting from the combined motions of epicycle and deferent. When the planet is outside the deferent, it appears dimmer and moves west to east; inside, it appears brighter because it is closer and at closest approach it moves east to west (retrograde)

THEORICAE NOVAE PLANETARVM GEORGII PVRBACHII ASTRONOMI CELEBRATISS.

DE SOLE

Ol habet tres orbes a se iuicē omniquaq; diuisos atq; sibi cōtiguos Quorū supre-mus secūdū superficiē conuexā est mūdo cōcentricus:secūdū cōcauā aūt eccētricus Infimus uero secūdū cōcauā cōcentric⁹: sed secūdū conuexā eccētric⁹ Tertius aūt i horū medio locatus tam secūdū super-ficiem suā conuexā q̄; concauā est mūdo eccentric⁹.Dicit̄ aūt mūdo cōcētric⁹or-

THEORICA ORBIVM SOLIS.

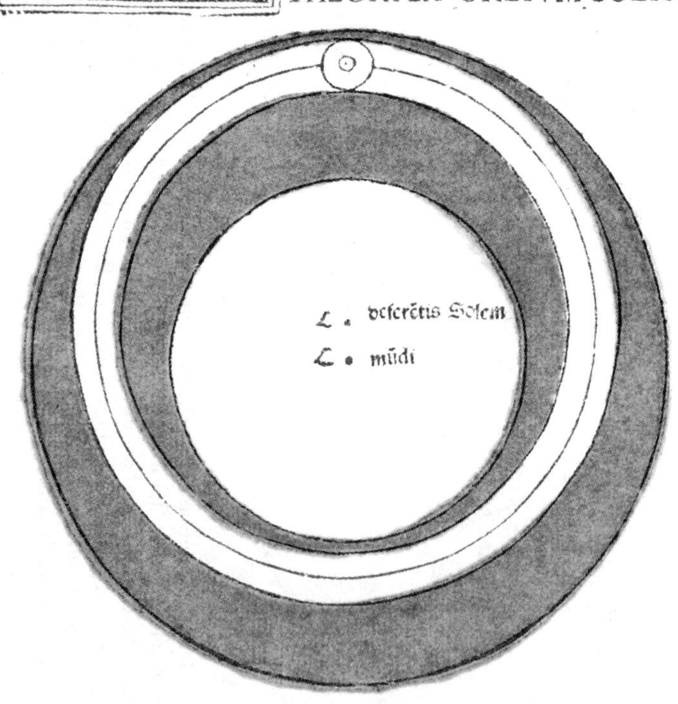

5. An adaptation of Ibn al-Haytham's thick-spheres model popularized by Georg Peurbach and included in 15th-century and later editions of the standard textbook of astronomy, Sacrobosco's *The Sphere* – this image from the 1488 Venice edition shows the Sun's sphere

off-centre; why doesn't it move to the centre? Why would heavy objects fall to something other than the centre? This discrepancy between the mathematical model and the physical system vexed medieval Arabic authors while both Aristotle's and Ptolemy's works were unknown in Europe. Ibn al-Haytham, or al-Hazen (c. 965–1040) adopted a compromise. His system had spheres centred on the Earth, which kept physicists happy. But these spheres were thick and solid enough to contain circular tunnels not centred on the Earth, through which the planets moved on their epicycles and deferents, which accounted for observations (Figure 5).

Medieval European astronomers inherited these ideas and problems and, like their Arabic colleagues, continued to refine and update the system, striving to maintain the best accuracy in predicting planetary positions or, somewhat less frequently, trying to generate a physically satisfactory system.

Early modern astronomical models

Nicholas Copernicus (1473–1543) spent most of his life as canon – an administrative post in Holy Orders – for the cathedral church in Frauenburg (today Frombork, Poland). He studied canon law in Bologna and medicine in Padua, and earned a doctorate in law at Ferrara in 1503. While at Bologna he began studying astronomy, and around 1514 he wrote an outline of his idea that the Sun, not the Earth, was at the centre of the planetary system. In his *heliocentric* (Sun-centred) system, the Earth rotates on its axis once a day, producing the familiar appearance of the entire cosmos turning around the Earth. What seems to be the motion of the Sun through the zodiac is really an illusion caused by the Earth's motion around the Sun. The observed 'loop' and retrograde motion of Mars, Jupiter, and Saturn does not result from their own motion, but rather from a combination of *ours and theirs* whenever the Earth laps one of these planets in the race around the Sun (Figure 6). Only the Moon revolves around the Earth.

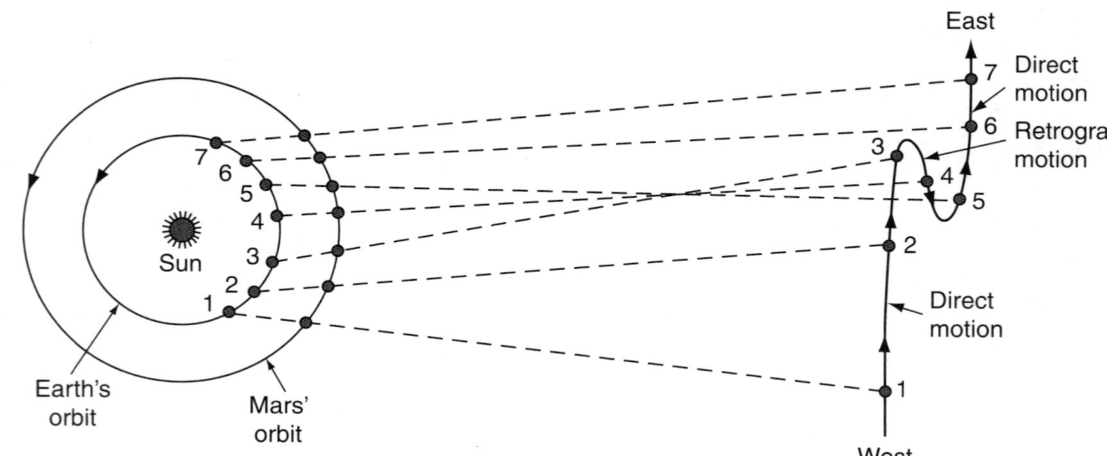

6. Copernicus's explanation of retrograde motion for one of the 'superior', or outer, planets (Mars, Jupiter, or Saturn). The 'loop' is an illusion caused when the Earth moves past one of these planets

Copernicus's work circulated in manuscript, and sufficiently established his reputation as an astronomer that in 1515, when a Church council was considering how to reform the old Julian calendar – in use since the Romans and now in need of an overhaul – they wrote to ask Copernicus's opinion. (Copernicus told them that the length of the solar year needed to be established more accurately first.) But Copernicus was reticent to publish a complete exposition of his system. He kept refining it for over 25 years, and had it not been for the nagging of several prominent churchmen it might never have been published. In 1533, for example, Johann Albrecht Widmannstetter, the Pope's personal secretary, lectured on Copernicus's system to the delight of Pope Clement VII and several cardinals. The cardinal of Capua, Nicolaus Schönberg, wrote to Copernicus saying that:

> I have learned that you teach that the Earth moves; that the Sun occupies the lowest, and thus the central place in the world . . . and that you have prepared expositions of this whole system of astronomy. . . . Therefore I most strongly beg that you communicate your discovery to scholars.

But Copernicus continued to demur, kept busy with his duties as canon, and expressing fear of criticism over his system's novelty.

In 1538, a young professor of astronomy named Georg Joachim Rheticus, sent from the University of Wittenberg by Melanchthon, came to study with Copernicus. Rheticus compiled and published a summary of Copernicus's ideas; the response was sufficiently positive that Copernicus finally agreed to publish his full manuscript, and gave it to Rheticus to shepherd through the press. Rheticus embarked on the task, but then took a job in Leipzig and handed off the project to a Lutheran minister named Andreas Osiander. Osiander finished the publication, and *On the Revolutions of the Heavenly Orbs* finally appeared in 1543 – a copy reaching Copernicus just before he died.

The book's appearance did not unleash the criticism Copernicus feared. It was read, but few readers were convinced. There were probably no more than about a dozen convinced Copernicans for the rest of the century. Why? Copernicus's heliocentric system did not fit observational data any better than the geocentric, nor was it physically much simpler. In fact, Copernicus had to keep using epicycles and an off-centre Sun in order to make his system harmonize with observations. More seriously, a moving Earth conflicted with basic physics, common sense, and possibly Scripture. Heavy bodies, like the Earth, naturally fall to the centre of the universe, its lowest point – this principle of 'natural place' explains why heavy objects fall. So how could the entire Earth remain suspended so far from the centre? Common sense indicates we are not moving. To rotate once a day, the Earth would have to turn very fast, yet we have no sensation of motion and neither birds in flight nor clouds are left behind by an Earth spinning beneath them. Some medieval thinkers had discussed the possibility of a rotating Earth. Nicole Oresme (c. 1325–82) concluded that all motions are relative, and without a point of reference it is impossible to decide whether the Earth is spinning or the heavens revolving. But it seems more likely, he concluded, that the Earth is stable and the heavens moving. More literal readers of Scripture could cite passages that speak of a stable Earth and a moving Sun, although interpretations varied widely. Finally, if the Earth moves around the Sun, the stars should exhibit parallax – a small shift in their apparent relative positions as the Earth swings from one side of its orbit to the other. But no parallax could be detected, meaning that either the Earth was *not* moving, or the stars were *incomprehensibly* far away. The 13th-century Campanus of Novara estimated that Saturn's sphere was about 73 million miles away – a staggering distance to even the best-travelled medievals – and the fixed stars lay just beyond. Copernicus estimated Saturn's sphere to be about 40 million miles away, but the lack of stellar parallax meant (according to later calculations) that the stars would have to be at least 150 *billion* miles farther out. This enormous gulf of emptiness seemed absurd to Copernicus's readers. (In fact, the closest star is 170 times farther away than the most modest prediction made from the lack of visible parallax. Stellar parallax was not detected until 1838.)

Several factors seem to have convinced Copernicus of heliocentrism, even without observational evidence. In his dedicatory letter to Pope Paul III, Copernicus referred to the Ptolemaic system, with its eccentrics, epicycles, and treatment of each planet separately, as a 'monster'. Noting that the world is 'created by the best and most systematic Artisan of all', it should be harmonious. Copernicus, as the humanist he was, saw himself as clearing away later 'accretions' to return to Plato's original challenge of showing the well-ordered nature of celestial motions. Worried about the 'novelty' of his system, he tried to minimize the appearance of novelty by citing ancient precedents – Aristarchus of Samos, Pythagoras, a certain Nicetas mentioned by Cicero – and even reinterpreting some Bible passages to favour heliocentrism.

One could, however, appreciate Copernicus's system without believing it to be true. Tables for determining planetary positions were easier to calculate in a heliocentric system; therefore, some astronomers adopted it as a 'convenient fiction'. Copernicus himself presented heliocentrism as a true description of the world, but Osiander undercut him by surreptitiously adding his own (unsigned) preface to Copernicus's book. Osiander wrote that we are 'absolutely ignorant of the true causes of planetary motions' and that:

> it is not necessary that these hypotheses be true or even probable; one thing suffices, that they give calculations matching observations . . . let no one expect anything certain from astronomy since it can give no such thing, nor should he take up anything confected for another purpose as if it were truth, lest he leave this study more stupid than he arrived.

Had Copernicus not already suffered a stroke, he might have had one when he saw Osiander's words. Rheticus was furious, and in his copy of the book scratched out Osiander's preface. The tension is once again between mathematical models and physical systems. Most astronomers were interested primarily in where planets would be when; whether the Sun went around the Earth or the Earth around the Sun simply did not matter, and many doubted whether one could ever tell for sure which was true. It was enough for an astronomical theory to provide tables and calculations to get planetary positions right. For the majority, practical results trumped theory. To understand this situation, we have to realize that the major driving force behind astronomical studies since before the time of Ptolemy was astrology, a practical enterprise that required being able to calculate planetary positions down to the minute, often many years in the past or the future.

Practical astronomy, or, astrology

Astronomy ('laws of the stars') measured and calculated the positions of heavenly bodies and hypothesized cosmological systems; astrology ('study of the stars', compare with geology, biology, and so on) endeavoured to explain and predict the heavenly bodies' effects on Earth. In general, these two endeavours – the first theoretical, the second practical – were pursued by the same people. Many early modern astronomers made their living primarily from practising astrology. Do not confuse ancient, medieval, or early modern astrology with the inanities of 'newspaper horoscopes'. Astrology was a serious and sophisticated practice based on the idea that heavenly bodies extend certain influences to earth – a key part of the conception of an interconnected world. Most medieval and early modern astrology is not 'magical', supernatural, or irrational; it depends upon natural mechanisms that are simply part of the way the world is put together. Light reaches us from the planets, so why shouldn't some additional influence accompany that light, just as the light from a fire also warms objects at a distance? Celestial influences on Earth are easy to observe – the Moon's link with the tides, or the Sun's zodiacal position with seasonal weather. Effects on the human body are clear as well, such as the synchronicity of the lunar cycle with menstruation. The reality of celestial influences seemed too obvious to question; the many controversies over astrology involved rather the extent of these influences and how to predict their effects accurately. The system of crisscrossing influences from seven planets constantly changing their positions relative to one another ('aspects'), and forever moving through twelve zodiacal signs which were themselves unceasingly passing through twelve 'houses' (positions relative to the horizon), made for an incredibly complex system. The complexity of indications and counterindications, knowns and unknowns, is comparable to the modern task of identifying factors in global climate change or predicting future economic trends. Relative to the latter enterprise, early modern astrologers possibly had a better success rate.

Astrology included several overlapping branches. Meteorological astrology endeavoured to predict the weather for the coming year. Many practitioners – often simply called 'mathematicians', a testimony to the calculations required for astrology – made a living producing almanacs containing calendars, lunar cycles, dates of eclipses, weather predictions (like the *Farmers' Almanac* today), and prognostications of important events or trends. The printing press made these publications inexpensive and widely distributed. Physicians used medical astrology to suggest crucial times for treatments and the course and possible causes of illnesses (see Chapter 5). Natal astrology used planetary positions at the exact time and place of a person's birth in order to determine what influences they 'imprinted' on the newborn. The specific combination of planetary influences would produce a unique 'complexion', or innate constitution, in the humoral system, leading to particular tendencies and traits. These tendencies (proneness to certain illnesses, to anger, laziness, melancholy, and so on) could be temporarily enhanced by subsequent planetary alignments. The goal of such astrology was thus to obtain information about a person's natural constitution, in order to be aware of particular strengths and weaknesses, and to provide advance notice of potentially dangerous or salubrious times. In stronger forms, this practice shaded into a judicial astrology that was criticized as unacceptably deterministic, namely, that astrological influences direct our actions and fates. Theologians condemned such notions as a violation of human free will. The scholarly consensus in the early modern period was that 'the stars incline but do not compel' us, and that *sapiens dominatur astris* ('the wise man rules the stars'). In short, human beings can always choose their actions, although the completely free exercise of will could be subject to external influences (such as a diminished capacity for reason owing to temporary irascibility due to a humoral imbalance caused by a particular position of Mars). Indeed, a parallel can be drawn between early modern astrology and current 'nature versus nurture' debates in their mutual attempts to explain human behaviour. The notable difference, ironically, is that moderns seem to have forgotten about the primacy of free will.

Some judicial astrology attempted to identify propitious dates for important endeavours. The mathematician and magus John Dee (1527–1608/9) used astrology to choose the best coronation day for Elizabeth I. A horoscope was cast for the founding of the Lincei, one of the first scientific societies, and also for the date of setting the cornerstone of the new St Peter's in Rome. Some astrological dates were chosen not to take advantage of a favourable 'influence', but rather to add levels of meaning to an event, the way for example, American scientists chose the landing date for probe on Mars to coincide with US Independence Day. Other forms of judicial astrology endeavoured to predict future events – such as wars and deaths – thus potentially moving away from the *natural* causality whereby learned early modern astrology was considered to operate. One way around this problem was to consider certain celestial events, comets in particular, not as *causes* but rather as *portents*, divinely sent signs of things to come. The interest in celestial portents was more pronounced in northern, that is, Protestant Europe, partly thanks to a preface written by Philipp Melanchthon for Protestant editions of Sacrobosco's *Sphere* – a fundamental astronomy textbook – in which he underscored the importance of astrology for reading God's signs in the heavens. In sum, astrology of various sorts was a source of helpful information for better living; its ubiquity in early modern thought emphasizes how the superlunar world was truly half of people's daily world.

Heavenly changes and divine harmonies

Astrological concerns over a heavenly portent led to the debut of the Danish astronomer and nobleman Tycho Brahe (1546–1601). In November 1572, he saw a bright object in the constellation Cassiopeia where none should be. Tycho was astonished – what could this object be, and what did it mean? In his astrological almanac for 1573, Tycho struggled to explain the object, concluding it was a divine portent of tumultuous changes to come. Tycho watched this brilliant point of light, but it did not move like a comet would. He and others around Europe tried to measure its diurnal parallax in order to determine its distance, but they found none, meaning that it was much farther away than the Moon – in the superlunar world, the realm thought free from change, yet it was a *new* star. (What Tycho saw was a supernova; the expanding remnants of that cataclysmic detonation were located in 1952. The term *nova* comes from Tycho's Latin for this object – *stella nova*, or new star.)

Soon after, in 1577, a bright comet appeared. Aristotle had taught that comets, like meteors, were sublunar phenomena caused by the ignition of flammable exhalations in the upper atmosphere. As erratic, changing objects, they could have no place in the unchanging superlunar world. Astrologically, Tycho concluded that the comet of 1577 continued the warning given by the new star, but this time he detected a diurnal parallax. His measurement, confirmed by others, indicated that the comet was far beyond the Moon, in the sphere of Venus. Tycho observed the same thing in 1585 when another bright comet appeared. These comets provided further evidence of change in the 'immutable' heavens, and their positions indicated that they were passing *through* the planetary spheres, implying that there were no solid spheres moving the planets. What then kept the planets in their regular courses? This puzzling liberation of the planets from solid spheres meant that the paths of celestial objects could cross one another, which in turn liberated Tycho to devise a new system of the heavens that combined his observations with the best parts of Ptolemy and Copernicus, while avoiding objectionable parts of both. In Tycho's *geoheliocentric* system, the Earth remained at rest at the centre, as common sense and Scripture dictated, with the Moon in orbit around it. The planets however all circled the Sun, which moved with its planetary retinue around the Earth.

While Tycho, within the castle-observatory Uraniborg he built on the island of Hven in the Danish sound, continued to observe and make the most accurate measurements of the heavens ever performed, Johannes Kepler (1571–1630), a convinced Copernican, was making his own astonishing discoveries on paper. In the 1590s, while teaching at a high school in Graz, Kepler puzzled over a question modern scientists would not think to ask. In Copernicus's system, there were only six planets orbiting the Sun, no longer seven orbiting the Earth. Seven planets had matched up nicely with the seven days of the week, the seven known metals, the seven tones of the musical scale, and all the other significant sevens in the world. Seven planets had a pleasing harmony, appropriate for an interconnected world; six did not. Why then were there six and only six planets, and why had God placed them at the distances He did? In the early modern world, a world full of meaning and purpose, everything has a message to be read.

While lecturing on 19 July 1595, Kepler suddenly realized that if one inscribes a regular polygon (triangle, square, pentagon, etc.) within a circle, and then inscribes a circle within that polygon, one obtains two circles whose relative sizes are fixed by the choice of polygon. Excitedly, he began calculating the ratios determined by different polygons to see if any of them matched up with the ratios of planetary distances from the Sun. They did not. Undaunted, he tried spheres and polyhedra instead of circles and polygons. In this case, by nesting spheres and polyhedra in the right order, Kepler obtained spheres whose sizes matched the distances of planets from the Sun estimated by Copernican theory. Moreover, since there exist only five regular polyhedra (that is, solid bodies where all the faces are identical, the five so-called Platonic solids: tetrahedron, cube, octahedron, dodecahedron, and eicosahedron) to use as spacers, there can be *six and only six* spheres, and thence, exactly *six* planets. For Kepler, this was an awesome discovery. He had found the cause for the number and distances of planets, and uncovered a geometrical structure to the heavens whose elegant beauty served as the best proof yet of the Copernican system. This striking correlation could not be random; Kepler had uncovered the mathematical blueprint by which God constructed the heavens.

Kepler exemplifies the unity of human inquiry normal for the early modern period. Theological and scientific inquiry are not separate: to study the physical world means to study God's creation, to study God means to learn about the world. Indeed, Kepler became convinced of Copernicanism partly because the heliocentric universe provided a physical analogy to the Holy Trinity: God the Father symbolized by the central Sun, God the Son by the sphere of the fixed stars that receives and reflects the Sun's rays, and God the Holy Spirit, theologically the love between Father and Son, by the light-filled space between the two. Drawing upon the idea of the Two Books, Kepler and his contemporaries were certain that God built messages into the fabric of creation for man to uncover. Thus theological motivations – the desire to read those messages in the Book of Nature – provided the single greatest driving force for scientific inquiry throughout the entire early modern period.

Kepler announced his discovery in the *Cosmographic Mystery* (1596) and sent a copy to Tycho Brahe. Tycho invited Kepler to collaborate; Kepler initially declined, but after Tycho moved to the court of Emperor Rudolf II in Prague as Imperial Counselor, Kepler joined him there in 1600. When the Danish nobleman died the next year, the Emperor made Kepler his Imperial Mathematician. Tycho had set Kepler studying the motions of Mars, and after a long struggle endeavouring to give it a path that fit the positions Tycho observed, Kepler came to a startling conclusion. He found that the planet's positions could be accounted for best by making it move on an *ellipse* not a circle. Kepler thus, reluctantly, broke with two thousand years of astronomical tradition focused on circles. But since (in Kepler's words) Tycho had 'smashed the crystalline spheres', what kept the planets moving in elliptical paths? Kepler postulated a 'moving soul' (*anima motrix*) in the Sun, a power that pushed the planets along. Like the Sun's light, this force decreased with distance, hence the further a planet was from the Sun, the more slowly it was moved. Drawing on the recent claim of William Gilbert (1544–1603) that the Earth is a gigantic magnet (see Chapter 4), Kepler postulated a second solar force, analogous to magnetism, that attracted the planets at some points and repelled them at others. The combination of *anima motrix* and 'magnetic' virtue kept the planets moving in ellipses, without the need for governing spheres, moving faster when they were pulled closer to the Sun and more slowly when they were pushed further away. Even as Kepler abandoned uniform circular motion, he was delighted to detect another uniformity to replace it: the 'equal area law', namely, that a line from the Sun to a planet sweeps out equal areas in equal times as the planet moves. Likewise, even as he helped dismantle Aristotle's cosmos, Kepler subtitled his *Epitome of Copernican Astronomy* as a 'supplement' to Aristotle's *On the Heavens*. Both continuity and change, both innovation and tradition, characterize early modern natural philosophy.

Telescopes and the Earth's motion

Tycho had been the greatest naked-eye observer; he was also among the last. While Kepler calculated, Galileo Galilei (1564–1642) heard of a Dutch device that made distant objects appear closer, built an improved one for himself, and turned it to the heavens in 1609. Virtually everywhere he directed his *occhiale* (later called the telescope) he made new discoveries. He found the Moon's surface covered with mountains, valleys, and oceans – in other words, looking just like the Earth and therefore made of the same four elements, not the Aristotelian quintessence. He found four moons orbiting Jupiter, like a little planetary system, and earned himself a patron and promotion by naming them the Medicean stars after Cosimo II de' Medici, Grand Duke of Tuscany. He found that Saturn had a strange shape that looked to him like three spheres joined together. He found that the planet Venus showed phases like the Moon. This last discovery was the first solid evidence against the Ptolemaic system, in which Venus could never be more than a crescent since it would always lie between the Sun and the Earth. Galileo's observation of both a crescent *and* a full Venus proved that it must sometimes be between us and the Sun and sometimes on the far side of the Sun, in short, that it orbits the Sun. Henceforth, astronomers would have to choose between versions of the Tychonic or the Copernican systems (Figure 7). The question of the mobility of the Earth – the only point that divided Tycho from Copernicus – thus came to be of prime concern.

Galileo rushed his first telescopic discoveries into print as the *Starry Messenger* (1610), sending them out to astronomers and rulers throughout Europe along with telescopes. Many had difficulty seeing what he described because the magnifications were low, the optics poor, and the telescope difficult to use. A key endorsement came from the Jesuit astronomers in Rome, who confirmed and continued Galileo's observations, and gave a feast

7. Three world systems compared in the emblematic frontispiece of Giovanni Battista Riccioli's *Almagestum novum* (Bologna, 1651). Astraea, goddess of justice, weighs the systems of Copernicus and Riccioli (a slight modification of Tycho's) while Ptolemy reclines alongside his now-discarded system. Above, cherubs carry the planets showing recent discoveries: the phases of Mercury and Venus, the Moon's rough surface, Jupiter's satellites, and Saturn's 'handles'. The divine hand blesses the world, its three extended fingers marked 'number, weight, measure' (Wisdom 11:20) expressing the mathematical order of creation

in his honor in 1611. The senior member of the Collegio Romano, Christoph Clavius (1538–1612), one of the most respected mathematicians in Europe and the man who had devised the new Gregorian calendar put into place in 1582 by Pope Gregory XIII (and still in use today), wrote that Galileo's discoveries required a rethinking of the structure of the heavens. Although Clavius and many others maintained geocentrism, some younger Jesuit astronomers were probably converted to heliocentrism. These excellent relations, however, would not weather Galileo's disputes (in which he often became insulting) with two Jesuit astronomers: Christoph Scheiner over the priority of discovery and nature of sunspots, and Orazio Grassi on comets (Grassi supported Tycho's measurements that comets were celestial bodies, while Galileo insisted they were sublunar optical illusions).

There is no episode in the history of science more subject to mythologizing and misunderstanding than 'Galileo and the Church'. The events resulted from a tangle of intellectual, political, and personal issues so intricate that historians are still unravelling them. It was *not* a simple matter of 'science versus religion'. Galileo had supporters and opponents both inside and outside Church hierarchy. The events are tied up with offended feelings, political intrigue, who was qualified to interpret Scripture, being at the wrong place at the wrong time, and being caught between Church factions. The final trigger was Galileo's 1632 publication of the *Dialogue on the Two Chief World Systems*, which compared Ptolemaic and Copernican systems, obviously choosing the latter as true, and the Earth as mobile. Galileo's chief evidence was his notion that the Earth's motion caused the tides; in this, he was famously wrong despite being right about the Earth's motion. The Church had no direct stake in which system was true; neither geocentrism nor Aristotelianism was ever Church dogma. But the Church did have a stake in biblical interpretation, and not only did a moving Earth have implications for interpretation, but Galileo had rather rashly dabbled in that matter in the early 1610s to support his ideas. This looseness with Scripture resembled the licence being taken contemporaneously by Protestants to reject traditional interpretations in favour of their own personal ones. As a result, in 1616 Galileo was told, and agreed, to treat heliocentrism and the Earth's motion hypothetically and not as literally true until there was demonstrable evidence. In 1624, Galileo got from his friend Maffeo Barberini, now Pope Urban VIII, permission to write the *Dialogue*, provided that Galileo include the Pope's methodological argument that a natural phenomenon (like the tides) might have several possible causes, some of which may be unknowable, and so we cannot assign it a single cause with absolute certainty. Galileo complied, but then put the argument only on the last page of the book, in the mouth of the character made to play the fool throughout. Galileo also 'neglected' to tell Urban about his 1616 agreement. When the book appeared (with the approval of the Vatican's licensers and censors) and these facts came to light, Urban became furious, feeling that he had been deceived and humiliated. To make matters worse, this petty annoyance materialized while Urban was overwhelmed by diplomatic negotiations regarding the on-going Thirty Years War, mounting criticism, attempts to depose him, and rumours of his impending death. The Inquisition worked out a plea bargain for Galileo that would have sent him home with a slap on the wrist, but the angry Pope refused to accept it – he insisted on making an example of Galileo. Galileo was ordered to abjure the Earth's motion, which he did, and his book was suppressed. Significantly, several cardinals, including Urban's nephew, refused to sign the sentence against Galileo. Galileo was never – folklore aside – condemned as a heretic, imprisoned, or chained.

Galileo ended up under house arrest at his villa in the Tuscan hills. There he continued to work and train students, and wrote perhaps his most important book, the *Two New Sciences*. The impact of his sentencing is difficult to assess. On the one hand, it made some natural philosophers reticent to express Copernican convictions. News of Galileo's condemnation caused René

Descartes (1596–1650), for example, to suppress a recently completed book that embraced heliocentrism. Those in Catholic Holy Orders, like the Jesuits, were now unable to support Copernicanism openly, and so embraced the Tychonic system or variations upon it (Figure 7), although sometimes with a wink and a grin. On the other hand, scientific inquiry, including in astronomy, continued in Italy and other Catholic countries, although sometimes skirting sensitive topics.

Following the conceptual upheavals of two previous generations, the mid-17th century witnessed more observational and technical developments in astronomy than theoretical ones. The French priest Pierre Gassendi (1592–1655) became the first person to witness a transit of Mercury across the Sun's disc in 1631; the event had been predicted by Kepler, who had died in 1630. Improved telescopes led to new discoveries and better measurements, but the need to avoid the distortions from spherical and chromatic aberration meant that telescopes had to be made longer and more unwieldy, sometimes over sixty feet in length. Nevertheless, the odd shape of Saturn was resolved into a ring system, and its largest moon discovered by Christiaan Huygens (1629–1695) in 1656. Gian Domenico Cassini (1625–1712), working in Paris and aided by the superior telescopes made by the Roman optician Giuseppe Campani, added four more, and named them Ludovican Stars after Louis XIV. The Jesuit Giovanni Battista Riccioli (1598–1671) produced a new star catalogue, and with his confrère Francesco Maria Grimaldi (1618–1663), a detailed lunar map providing many of the names still used today for its features – including naming one of the most prominent craters after Copernicus. In Gdansk, Johann Hevelius (1611–87), probably the last person to make careful measurements with both the telescope and the naked eye, also prepared a lunar map, as well as observing comets and participating in the Europe-wide discussion of whether their motion was rectilinear or curved into an orbit around the Sun.

The problem of how the planets keep moving in constant orbits without the aid of solid spheres continued to attract speculation. Descartes proposed a comprehensive world system that became one of the most important of the 17th century. He envisioned all space to be filled with invisibly small particles of matter. These particles moved always in circular streams or vortices. Our solar system was a gigantic vortex of these particles that carried the planets along like a whirlpool carries along bits of straw. This vortex model neatly explained why the planets all move in the same direction and nearly in the same plane. The Earth itself lay at the centre of a smaller vortex that kept the Moon moving in its orbit, and the swirling of matter around the Earth provided a 'wind' that pushed objects toward the centre of the Earth, thus producing the phenomenon of gravity. Descartes' vortex theory gave a comprehensible explanation for celestial motions, and it was widely disseminated in popular treatments and textbooks, but it remained too imprecise to be of practical value for astronomers.

A young Isaac Newton (1643–1727) embraced the Cartesian vortex theory. As a student at Cambridge in the early 1660s, Newton studied the Aristotelian works that remained standard undergraduate texts in most universities. But he soon began an extracurricular reading of the ideas of the 'moderns' such as Descartes. He adopted a modified version of Descartes' principles for both planetary motions and gravity. But by the early 1680s, Newton had begun to think differently. He discarded Cartesian vortices and began thinking in terms of an attractive force existing between Sun and planets. He had at his disposal several sources for this idea, most notably the familiar phenomena of magnetism and the 'magnetism-like' force between Sun and planets that Kepler postulated. For Kepler, the combination of the *anima motrix* and this 'magnetism' produced the planets' elliptical orbits. For Newton, it would be the balance between inertia (the tendency of the planet to move in a straight line tangent to its orbit) and the force of attraction (what we call gravitation) towards the Sun that produced stable elliptical orbits. Several members of the Royal

Society of London had been working along similar lines to explain planetary motion, most notably Robert Hooke (1635–1703), who wrote about his ideas to Newton in 1679–80. Hooke's subsequent complaint that Newton had taken his idea without giving him sufficient credit led the neurotically hypersensitive Newton to expunge any reference to Hooke from his writings and to treat him antagonistically for the rest of his life. Newton's great achievement, published in the *Mathematical Principles of Natural Philosophy* (1687), was to rederive purely mathematically the laws of planetary motion that Kepler had derived empirically from Tycho's observations, and to make gravitation truly universal – that is, existing mutually between all parcels of matter. Kepler would undoubtedly have been pleased; here was more evidence of the harmonious mathematical plan upon which God had created the world. Newton's law of universal gravitation obliterated the last traces of the former distinction between terrestrial and celestial physics – the same law governed the revolution of the planets and the fall of an apple.

Not everyone was pleased. By reviving the idea of attractive forces, Newton seemed to be resuscitating an idea unpopular for some 70 years. An invisible, immaterial force without mechanism or identifiable cause that operated between all bodies was not only less comprehensible than material Cartesian vortices, but seemed to many as a return to the 'hidden qualities' of Aristotelians or the sympathies of natural magic. Indeed, the cutting edge of natural philosophy in the second half of the 17th century had been endeavouring to explain what appeared to be attractions and sympathies by means of a mechanical exchange of invisible particles (see Chapter 5); now Newton seemed to be turning back the clock.

Gottfried Wilhelm Leibniz (1646–1716), with whom Newton waged a priority dispute over the invention of calculus, accused Newton's 'hidden attractive quality' of 'confounding the principles of true philosophy' and returning it 'to the old asylums of ignorance'. While Newton's apologists asserted that gravitational attraction was simply a fundamental property of matter, Newton himself did want to find gravity's *cause*. His method of pursuing that answer, however, reminds us that Newton was not some 'modern scientist' accidentally born in the 17th century. Newton, perhaps with uncharacteristic modesty, considered himself to be only the rediscoverer of the law of universal gravitation; it had been known to the ancients. For Newton believed in the *prisca sapientia*, an idea popular among many Renaissance humanists of an 'original wisdom' divinely revealed aeons ago and corrupted over time. He strove to interpret Greek myths, biblical passages, and the *Hermetica* to show that they concealed ideas about the hidden structure of the world, including his own inverse-square law of gravity. Newton seems to have thought – and believed the 'ancients' did as well – that gravitational attraction resulted from the direct and continuous action of God in the world. Like Kepler, who felt he had revealed God's geometrical blueprint, Newton considered himself chosen to restore ancient knowledge – and not just scientific knowledge. He spent years in theological and historical studies, believing that Christianity, like all other knowledge, had become corrupted over time, and endeavoured to restore its supposedly 'original' theology that did not include, for example, the divinity of Christ. He likewise laboured on ancient chronology, in part to get reliable reckoning dates for interpreting biblical prophecies about the end of the world. We return once again here to the broader, more inclusive view of natural philosophy relative to that of modern science. Newton saw 'the task of natural philosophy as the restoration of the knowledge of the complete system of the cosmos, including God as the creator and as the ever-present Agent'.

Chapter 4
The sublunar world

While many early modern natural philosophers looked up towards the heavens, even more looked anew at things on Earth. The sublunar world was the realm of the Earth and its four elements – earth, water, air, and fire – and the realm of change, of coming-to-be and of passing-away, a dynamic world of unceasing transformations. Heavy elements (earth and water) and heavy objects fell naturally towards the lowest point of the universe – its centre – where the Earth remained at rest. The light elements (air and fire) moved upwards towards the sphere of the Moon, the uppermost limit for the four elements. Thus each element found its 'natural place' in the scheme of things by means of a 'natural motion' based on its weight or levity. This Aristotelian system explained why rocks and rain fall downwards while smoke rises and a candle flame always points upwards. In the superlunar world, on the contrary, heavenly bodies were composed of the quintessence which, being neither heavy nor light, moved neither up nor down, but with eternal circular motion around the Earth. Early moderns re-examined the Earth, the elements, and the processes of change and motion, and formulated a range of systems for making sense of things. Some were expressly intended to replace the Aristotelian worldview, others tried only to refine it, and virtually none was completely free of Aristotle's influence. The result of observing, experimenting, and reconceptualizing the sublunar world was not the gradual formulation of a single worldview leading to the modern scientific perspective, but rather the creation of competing world systems that jostled for recognition and pre-eminence throughout the 17th century.

The Earth

Early modern natural philosophers considered the Earth, like the rest of the cosmos, to be only a few thousand years old. The chronology provided by the Bible, the oldest text available, drew the lineage of mankind back about 6,000 years. While only some readers interpreted Genesis 1 to describe a literal chronology involving six 24-hour days of creation (St Augustine had rejected such literalism in the 5th century), no one seriously thought that the Earth's prehuman history extended much further back in time. The largest estimates suggested a creation about 10,000 years old. This position was not a matter of dogma; there was simply no evidence to make one think otherwise. It was in the work of Niels Stensen (1638–86), better known by his Latinized name Nicholas Steno, that the idea of geological history emerged. Born in Denmark, Steno first applied himself to anatomy, becoming famous for his skill in dissection, with which he made important discoveries such as the salivary passage, known today as Stensen's duct. Like many other natural philosophers of his day, he toured European centres of learning, meeting other natural philosophers and exchanging new knowledge. In the 1660s, he settled in Florence under Medici patronage, and became interested in the layers of rock – what we call strata – visible in the Tuscan hills and the seashells found encased in them. He concluded that these layers must once have been soft mud laid down gradually by sedimentation, and therefore that lower strata must be older than higher ones. Wherever strata were not horizontal, he argued, they must have been disrupted by some upheaval after they hardened into stone. These conclusions did not cause Steno to revise estimates of the age of the Earth upwards – after all, mud can harden into brick relatively quickly – but they did indicate that the Earth's surface was subject to dramatic changes and that rocks preserve a record of these changes.

Towards century's end, several authors – especially in England – built upon Steno's work to compile 'histories of the Earth' to explain its current appearance. Most of them invoked global catastrophes as causal agencies and interleaved biblical and other historical accounts with natural philosophical ideas and observations. Thomas Burnet's *Sacred History of the Earth* (1680s) proposed six geological ages punctuated by cataclysmic biblical events. Edmond Halley and William Whiston (1667–1752), both associates of Newton, suggested that comets colliding with the Earth were major crafters of its history, causing such things as the inclination of the Earth's axis and Noah's flood.

Changes to the Earth's surface were studied first-hand by the Jesuit polymath Athanasius Kircher. While in Sicily in 1638, Kircher witnessed a violent earthquake and the eruption of Mt Etna. Vulcanism had not previously been a subject of study, in large part because Mt Vesuvius, the only active volcano on the European mainland, had been dormant for over three hundred years prior to its sudden and deadly eruption in 1631. Kircher travelled to observe the continuing eruption, and actually descended into the active crater to get a better look. He observed how volcanic action both destroyed old mountains and built new ones, dramatically altering the landscape. He attributed volcanic heat to the inflammation of sulphur, bitumen, and niter (a combination close to that of gunpowder) underground. Noting that the quantity of fire and molten rock emitted was too great to be produced within the mountain itself, he surmised that volcanoes must be vents for immense fires deep within the Earth. He thus concluded that the Earth cannot be merely 'pressed together from clay and mud after the Flood, hardly different from some lump of cheese', but had instead a complex and dynamic internal structure. He envisioned Earth's interior riddled with passages and chambers (Figure 8).

8. An idealized depiction of the hidden interior of the Earth and its volcanoes as envisioned by Athanasius Kircher, *Mundus subterraneus* (Amsterdam, 1665)

Some conveyed fire to volcanic vents from a fiery central core (he *never* literally conflated this core with Hell), while others allowed the passage of water, often from one sea to another. The flow of massive amounts of water through such passages generated ocean currents and turbulence. Collecting data from many sources, especially reports sent from Jesuit missionaries, Kircher compiled his encyclopedic *Subterranean World* (1665) containing, among much else, world maps showing ocean currents, volcanoes, and the possible locations of submarine passages.

In contrast to Kircher's observation of the Earth's most dramatic events, William Gilbert (1544–1603) performed quiet experiments at home to uncover another invisible feature of our planet. Gilbert, a physician to Elizabeth I, studied that ever-puzzling object, the magnet. His book *On the Magnet* (1600) surveys the properties of magnets, recounts experiments with them, and distinguishes magnetic attraction from the temporary ability of rubbed amber to attract straw. (For this latter phenomenon, he coined the word *electrical* – from the Greek *ēlectron* for amber.) Some of his experiments were inspired by those performed by Pierre de Maricourt in the 1260s, but Gilbert directed his studies towards a new goal. Pierre had used spherical magnets, or lodestones – pieces of the naturally magnetic mineral magnetite – and discovered that magnets have poles that he named north and south. Gilbert, also using spherical magnets, observed that iron needles placed on them mimicked exactly the behaviour of compass needles on the Earth. He thus concluded that the Earth is a gigantic magnet. It too has magnetic poles that attract the compass needle, just like a lodestone. (Previously it was thought that compasses pointed toward the *celestial* North Pole not towards a terrestrial pole.) In short, Gilbert used his spherical lodestone as a *model* of the Earth – reasoning by analogy, he extrapolated what he saw while experimenting with the lodestone to the whole Earth.

Gilbert's goal was to undergird Copernicanism, which had thrown the whole concept of natural place and natural motion into confusion. Putting the Earth in motion, spinning giddily on its axis and orbiting far above the centre of the universe, raised serious problems for physics. Why would heavy bodies fall to an Earth that is not at the centre? What caused the Earth to spin? Supporters of Copernicanism had to find a new physics that could reorder this chaos. Once Gilbert had argued that the Earth has magnetic poles, he stressed that those poles defined a real *physical* axis, and using the principle that everything in nature has a purpose, he argued that the purpose of that axis was to provide for the Earth's rotation. Furthermore, Earth's magnetic virtue animates it with intrinsic motive power; just as lodestones cause iron objects to move. The Earth's magnetic 'soul', as Gilbert calls it, not only causes compasses to turn north, but the planet to turn on its axis. Upon this foundation, Gilbert formulated a 'magnetical philosophy' wherein magnetic virtues permeate and govern the universe. Drawing upon the principle that like attracts like – the 'sympathy' of natural magic – the magnetic philosophy tried to solve the disruption of 'natural place' by suggesting that pieces of Earth are naturally attracted to the Earth, while pieces of the Moon are naturally attracted to the Moon. Thus earthly objects would fall towards the Earth regardless of the Earth's place in the cosmos. In Gilbert's universe, magnetic forces maintain order in both sub- and superlunar worlds, and his vision deeply influenced Kepler, Newton, and others.

Motion on Earth

While the magnetic philosophy tried to explain *why* bodies fall, Galileo endeavoured to describe mathematically *how* they fall. He built inclined planes, pendula, and other devices to study terrestrial motion. His *Two New Sciences* (1638), written while under house arrest, was the culmination of a study of motion he began in the 1590s. He discovered, contrary to Aristotle's claim, that all bodies fall at the same rate regardless of weight. With elegant logic he argued that if a ball rolled down an inclined plane speeds up and one rolled up an inclined plane slows down, then one rolled on a level surface – neither up nor down – would maintain a constant speed. Since on Earth that 'level' surface would actually be the curved surface of the globe, a ball rolled on its perfectly polished surface would circle it for ever. Using this 'thought experiment', Galileo both enunciated a principle of inertia (that moving bodies keep moving unless acted upon by an external agent), and brought the eternal circular motion of the heavens down to Earth – further eroding the distinction between sublunar and superlunar realms.

Methodologically, what Galileo ignored is as important as what he paid attention to. In describing motion, he never concerned himself with *what* is moving – a ball, an anvil, or a cow. In short, he ignored the *qualities* of bodies that Aristotelian physics emphasized. Galileo favoured instead their *quantities*, their mathematically abstractable properties. By stripping away an object's characteristics of shape, colour, and composition, Galileo gave idealized mathematical descriptions of its behaviour. A cold brown ball of oak doesn't fall any differently than a hot white cube of tin; Galileo reduces both objects to abstract, decontextualized entities able to be treated mathematically. A group known as the Oxford Calculators had begun applying mathematics to motion in the 1300s; in fact, Galileo begins his exposition of kinematics in the *Two New Sciences* with a theorem they enunciated.

But Galileo went much further by linking mathematical abstraction tightly with experimental observation. As he conducted innumerable experiments, he sifted out air resistance and friction as 'imperfections' from the ideal mathematical behaviour that can be experienced only in thought. Plato, with his idea of a world that only imperfectly follows the eternal mathematical patterns according to which it was fashioned, might have found something to agree with in Galileo's perspective (even if Aristotle would have objected). Evoking the Christian image of the 'Book of Nature', Galileo wrote famously that 'this grand book, I mean the universe . . . is written in the language of mathematics, and its characters are triangles, circles, and other geometrical figures, without which it is humanly impossible to understand a single word of it'. The technique of reducing the physical world into mathematical abstractions, and eventually into formulas and algorithms, championed by Galileo, played a key role in producing a new physics, and stands as a distinctive feature of the Scientific Revolution.

Significantly, Galileo is content to *describe* motion mathematically without worrying about its *cause*. This feature of Galileo's work departs fundamentally from Aristotelian science where true knowledge is the knowledge of causes. Galileo's approach resembles an engineer's – a person more interested in describing and utilizing *what* an object does than *why*. Here Galileo draws upon his Northern Italian context where engineering and the learned engineer had achieved great prominence (see Chapter 6). The *Two New Sciences* makes the importance of practical engineering clear: its interlocutors meet amid construction works in Venice's shipyards, and discuss beam and tensile strength and scale-ups and scale-downs – topics of critical importance to engineers and architects. As a young professor in Padua, Galileo supplemented his meagre university salary by tutoring on mechanics and fortification. His later study of projectile motion – showing that projectiles follow a parabolic path – which we tend to remember primarily as a contribution to the physics of motion, continued earlier studies by Niccolò Tartaglia (1499–1557), a learned engineer who wrote his own *New Science* in 1537 about applying mathematics to motion, especially the motion of cannonballs, a topic of immediate practical importance for Italy's ever-warring states. It is easy to make the development of science too abstract and cerebral, and to forget that it is often driven by pressing and very practical concerns.

Water and air

The study of water for engineering purposes led to a sequence of important discoveries by Galileo's followers. His student and successor to his mathematics chair at Pisa, the Benedictine priest Benedetto Castelli (1577–1643), devoted himself to hydraulics and fluid dynamics – important practical questions in an era when Italy was awash with grand waterworks projects involving canals, fountains, irrigation, aqueducts, and sewers. The need to move water greater distances vertically (for example, out of deep wells or mines) led to the discovery that siphons could not draw water upwards to a height of more than about 34 feet. In the early 1640s, Gasparo Berti (c. 1600–43), a colleague of Castelli's at the University of Rome, tried an experiment to study this problem. With co-workers including Athanasius Kircher, he took a pipe 36 feet long and able to be closed at both ends, and mounted it vertically with its lower end in a basin of water

(Figure 9, left). He closed the bottom valve and filled the pipe completely with water. Then he closed the pipe at the top and opened it at the bottom. The water began to flow out, but stopped suddenly when the height of the column of water left in the pipe fell to 34 feet. What kept the water suspended at 34 feet – no higher and no lower?

Castelli's student Evangelista Torricelli (1608–47), who would later be given Galileo's position as mathematician and philosopher to the court of Ferdinando II de' Medici, devised a simple instrument analogous to Berti's pipe, but easier to handle. He took a glass tube about a yard long, sealed it at one end, and filled it with mercury. When the open end was inverted into a basin of mercury (Figure 9, right), the mercury in the tube began to drain out, but stopped when the column of mercury remaining in the tube was about 30 inches in height, about one-fourteenth the height at which the water had stopped in Berti's pipe. Significantly, mercury is about 14 times as dense as water – meaning that the height of any fluid left suspended in a tube was a direct function of the fluid's density. Drawing upon ideas of fluid equilibria worked out in earlier studies of water, Torricelli explained these results by saying that the weight of fluid left in the tube was balanced by the weight of the external air pressing down on the fluid in the basin. The idea that air had weight conflicted with Aristotle's system where it had none. Torricelli proposed not only that we 'live submerged at the bottom of a vast ocean of elemental air', but also that his instrument could measure and monitor changes in the weight of that air, leading to a new name for his device: the *barometer*, literally the 'measurer of weight'.

Some of the most celebrated experiments of the 17th century were designed to explore ideas provoked by Torricelli's tube. An elegant experiment to prove that it is the atmosphere's weight that keeps liquids suspended in the tube was proposed by the mathematician and theologian Blaise Pascal (1623–62), and carried out by his brother-in-law Florin Périer in 1647. Following

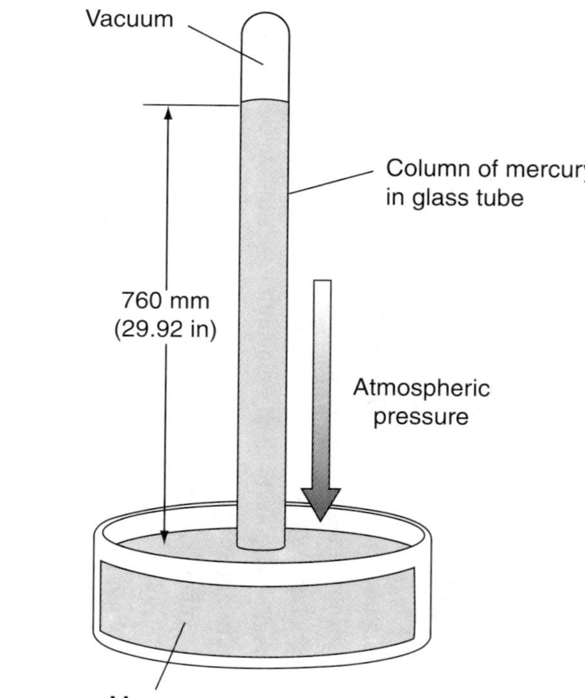

9. **(left) Gasparo Berti's water barometer depicted in Gaspar Schott, *Technica curiosa* (Nuremberg, 1664); (right) A schematic of Evangelista Torricelli's simplified mercury barometer**

Pascal's instructions, Périer prepared 'Torricellian tubes' in a monastery garden at the base of the Puy-de-Dôme, a mountain near their home in central France. He then carried one tube more than 3,000 feet up the mountain, where he found that the level of the mercury stood three inches lower. Upon returning down the mountain, the mercury regained its original height. At higher elevations, with less of the 'ocean of air' pressing down from above, the weight of air resting upon the mercury was reduced, and could therefore counterbalance less mercury in the tube.

A spectacular experiment performed before many spectators was that of the famed 'Magdeburg sphere' created by Otto von Guericke (1602–86), natural philosopher, mayor of Magdeburg, showman, and maker of wondrous devices. Von Guericke built two hemispherical copper shells with rims that fit smoothly together. He put them together to form a sphere, opened a valve installed on one half, and – using a device of his own invention modelled on a water pump – pumped the air out of the sphere. He closed the valve, and showed that teams of horses could then not separate the two halves because of the air's weight holding them together (Figure 10). Upon opening the valve, air rushed in, and von Guericke then easily separated the two halves with a flick of the wrist.

But what was in the space above the mercury or within von Guericke's sphere? Many experimenters believed it was literally *empty*, a vacuum – a highly controversial topic in the 17th century. Aristotelians and some others argued that a vacuum was impossible – as summarized in their catchphrase 'nature abhors a vacuum'. They saw the world as completely full of matter, a *plenum* – and some natural phenomena seemed to support them. They argued that the space contained air or some finer aerial substance drawn out of the mercury. Experiments endeavoured to resolve the point, but did not entirely settle the dispute between 'vacuists' and 'plenists'. Sound was not transmitted through the space, indicating that the air needed to carry sound had been removed. Yet light passed through – did not light, like sound, need

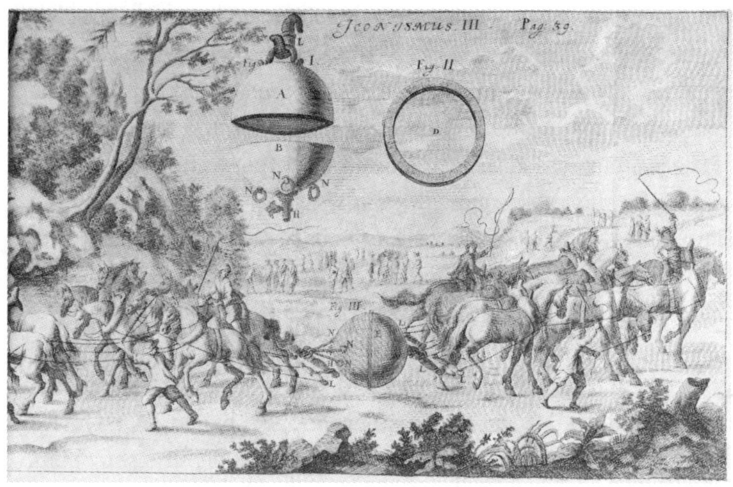

10. Otto von Guericke's showy demonstration that teams of horses could not pull apart the halves of a hollow sphere out of which the air had been pumped – evidence of the power of atmospheric pressure. Depicted in Gaspar Schott, *Technica curiosa* (Nuremberg, 1664)

some medium to transmit it? What are routinely seen as 'landmark' experiments in the history of science rarely proved as convincing to their contemporaries as they seem to moderns in retrospect. Experimenting, and especially interpreting results, is a tricky and contentious business, has always been and will always be so.

Robert Boyle (1627–91) soon joined the ranks of those studying air. As the youngest son of the richest man in Britain, Boyle had both time and resources to spend his life experimenting, mostly in his sister's house on London's Pall Mall, where he lived much of his adult life. He and several contemporaries noted the compressibility of air, specifically that the greater the pressure on a sample of air, the smaller its volume, a relation later called 'Boyle's Law' and still taught as such to chemistry students. In 1658, having heard of von Guericke's air pump, Boyle and the ingenious Robert Hooke built an improved version able to evacuate a large glass sphere, allowing various objects to be sealed up and observed as the air was pumped out (Figure 11).

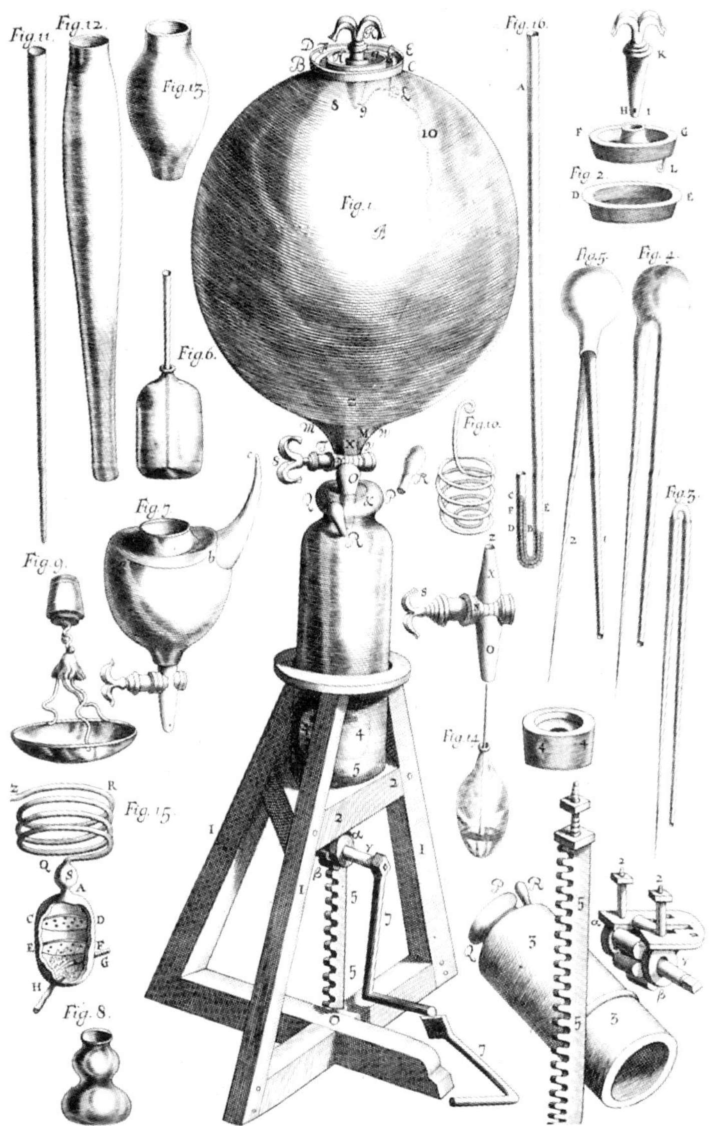

11. Robert Boyle's and Robert Hooke's air pump. Robert Boyle, *New Experiments Physico-Mechanicall Touching the Spring of the Air* (Oxford, 1660)

Boyle sealed a barometer (he probably coined this name for Torricelli's tube) in his air pump and watched the level of mercury drop as the air was withdrawn. He performed a dizzying array of experiments in the pump: from trying to ignite gunpowder, fire a pistol, or hear a watch ticking, to measuring how long various living creatures – cats, mice, birds, frogs, bees, caterpillars, and almost everything else – could survive without air. He also experimented with burning candles in the air pump, and noted the dependence of fire on the quantity of air available.

Fire: the chymists' tool

Long before the early modern period, the status of fire as a material element had been disputed. Amid such debates, one group regularly employed fire as their primary tool for studying and controlling matter and its transformations: the alchemists. The Scientific Revolution was alchemy's golden age. Today 'alchemy' is often taken to mean a single-minded (and futile) quest for making gold, something more or less 'magical' and thus distinct from chemistry. But in the early modern period, 'alchemy' and 'chemistry' referred to the same array of pursuits. Some historians today use the archaically spelled *chymistry* to refer to all these undifferentiated pursuits together. Gold-making, or *chrysopoeia*, was a key part of chymistry, but there was nothing 'magical' (in the modern sense) involved, simply a practice based on theories different from our own. Notebooks survive that record the daily operations of 'alchemists' and often reveal careful methodologies of experimental practice, textual interpretation, observation, and theory formulation. Besides the quest for gold, chymistry also included the broader study of matter and the production of articles of commerce such as pharmaceuticals, dyes, pigments, glass, salts, perfumes, and oils. The union of material production and natural philosophical speculation forms a central characteristic of this subject from its 4th-century origins in Hellenistic Egypt down to present-day chemistry.

The search for a method to turn lead into gold was not just wishful thinking. It rests upon the theory that metals are compound bodies produced underground by the combination of two ingredients called 'Mercury' and 'Sulphur'. When the two combine in the correct proportions and purity, they form gold. If there is not enough Sulphur, silver results. Too much Sulphur (a dry, flammable principle) produces iron or copper – their excess Sulphur can be demonstrated by the flammability, hardness, and the difficulty of melting these metals. Excess Mercury (a fluid principle) gives tin or lead – the soft, easily fusible metals. Thus transmutation was, in theory, a simple matter of adjusting these two components to the proportion found in gold. The observation that silver ores contained some gold and that lead ores contained some silver suggested that transmutation was occurring naturally underground, as poorly compounded metals were purified or 'matured' into more stable, better-concocted ones. The challenge was to effect this transformation artificially and faster. Chrysopoeians thus sought to prepare what they called the Philosophers' Stone, a material agent for bringing about transmutation. Once prepared in the laboratory, a small quantity of the Stone mixed with molten base metal was supposed to convert it into gold in a few minutes. Many texts claimed success in this process, and seekers after transmutation strove to replicate it. The difficulty lay in the intentional secrecy of such writings – the ingredients, process, and even the theory were hidden beneath codes, cover names, metaphors, and pictorial emblems, often of bizarre character (Figure 12).

Alchemy's secrecy arose in part from artisanal practices wherein it was necessary to preserve proprietary rights as trade secrets. Secrecy was encouraged by medieval laws forbidding transmutation out of fear of debasing the currency. But authors also justified secrecy by claiming that their knowledge was not only dangerous in the wrong hands, but also a privileged knowledge not to be divulged to those unworthy of it.

12. An alchemical allegory depicting the purification of gold and silver, a first step in making the Philosophers' Stone. The king represents gold, while the wolf jumping over the crucible (a vessel for refining metals) stands for the mineral stibnite, a material capable of reacting with and removing the silver and copper commonly alloyed with gold. The queen represents silver, and the old man (Saturn) lead, in reference to the process of cupellation, which uses lead to purify silver. From *Musaeum hermeticum* (Frankfurt, 1678)

The continuing British usage of 'chemist' to mean 'pharmacist' originated in the early modern period when most chymists devoted at least some of their efforts to making medicines. The application of chymistry to medicine began with the Provençal Franciscan friar Jean of Rupescissa (1310–c. 1362), who advocated the use of alcohol distilled from wine to prepare medicinal extracts. The use of chymistry to prepare medicinal substances expanded throughout the next century, before receiving its most vocal advocate in the larger-than-life figure of Theophrastus von Hohenheim, known as Paracelsus (1493–1541). Paracelsus criticized traditional medicine based on Greek, Roman, and Arabic authors and devised his own system based on a range of sources from direct observation to Germanic folk beliefs. He championed chymistry as the means to prepare virtually any substance into a powerful medicine, and showed little interest in chrysopoeia. His guiding idea was that noxious properties arise from impurities in otherwise wholesome substances, much like sin and death contaminated a world which, as God's creation, was intrinsically good. Using distillation, fermentation, and other laboratory operations, chymistry provided methods for dividing good from bad, medicine from poison. Paracelsus also taught that all substances were composed of three primary ingredients – Mercury, Sulphur, and Salt – a terrestrial trinity called the *tria prima* that mirrored the Divine Trinity and the triune nature of man – body, soul, and spirit. A process he called *spagyria* endeavoured to divide a substance into its *tria prima*, purify each, and then recombine them into an 'exalted' form of the original substance with enhanced medicinal power and no toxicity.

But Paracelsus went further: chymistry was not just a tool for making medicines, it was the key to understanding the universe. As Paracelsus' late 16th-century followers systematized his often chaotic writings (which, it was rumoured, he dictated when drunk), they formulated a chymical worldview that envisioned virtually everything as fundamentally chymical. The cycle of rain through sea, air, and land was a great distillation. The formation of minerals underground, the growth of plants, the generation of life forms, as well as the bodily functions of digestion, nutrition, respiration, and excretion were all seen as inherently chymical. God Himself became not the geometer of the Platonists, but the Master Chymist. His creation of an ordered world out of primordial chaos was akin to the chymist's extraction, purification, and elaboration of common materials into chymical products, and His final judgement of the world by fire like the chymist using fire to purge impurities from precious metals. This worldview saw even man's ultimate destiny as chymical. Upon death, the human soul and spirit separate from the body. The material body putrefies in the grave until, at the general resurrection, it is renewed and transformed, whereupon the purified soul and spirit are reinfused by God the Chymist to produce a glorified and eternal human being, just as in spagyria the *tria prima* are separated from a substance, purified, and recombined into a resynthesized and 'glorified' product.

Paracelsianism attracted many adherents. When Tycho first saw his nova in 1572, he had just stepped out of a laboratory where he was preparing Paracelsian remedies. He later built a laboratory into his observatory-castle in order to study what he called 'terrestrial astronomy', namely, chymistry ('as above, so below'). Because of the anti-establishment nature of Paracelsus' style, often expressed in rants against the Classical learning, universities, and licensed physicians, his ideas provoked heated debate and often found their greatest following among those outside of established circles. Indeed, chymistry as a whole lived most of its existence outside traditional halls of learning and suffered from an uneasy status. While physics and astronomy formed required parts of university study from the Middle Ages on, chymistry did not obtain an academic footing until the 18th century. One reason is that it could boast no Classical roots; neither Aristotle nor any other ancient authority wrote about it, unlike astronomy, physics, medicine, and the life sciences. Its close link to commerce and artisanal production, its practicality – and often messy, laborious, smelly character – further disabled it from being considered among respectable topics. Yet chymistry's emphasis on practical experiment also meant that it amassed a huge inventory of materials, knowledge of their properties, and facility for working with them. The commercial importance of this knowledge increased substantially throughout the 17th century and many chymists took an entrepreneurial route – sometimes engaged by princely or other patrons and mining operations to improve yields or seek transmutation, sometimes working independently to introduce new wares to the market place. Unfortunately, chymistry's ability to imitate gems and metals, and the claim of chrysopoeia to make gold, provided opportunities for fraud, leading to a widespread connection of chymistry with unscrupulous practices. Already in the late Middle Ages, Dante had put chymists – 'the apes of Nature' – into the eighth circle of Hell alongside counterfeiters and forgers, and later, 17th-century playwrights such as Ben Jonson in his *Alchemist* (1610) used the figure of the false chymist and his greedy clients to comic effect.

Most 17th-century training in chymistry took place in medical contexts. In Germany, Johannes Hartmann (1568–1631) became the first professor of *chemiatria* (chymical medicine) in 1609. His appointment was made at the University of Marburg, a Calvinist institution newly established (and hence more able to be innovative) by Moritz of Hessen-Kassel, a prince whose court supported chrysopoeians, Paracelsians, and other chymists. In France, regular chymistry instruction began at the Jardin du Roi in Paris, a botanical garden founded to propagate and study medicinal plants. A succession of lecturers at the Jardin gave 'how to' courses based on laboratory demonstrations that were open to the public. Private lecturers, often pharmacists, also offered courses of chymistry, such as Nicolas Lemery who taught from his house in Paris. His textbook *Cours de chymie* (1675) became a best-seller. Indeed, the dozens of chymical textbooks published in France and Germany established a didactic tradition that compensated for chymistry's absence from university curricula.

Chymistry's practical flavour does not mean that it did not contribute significantly to natural philosophical theories – quite the opposite. One of the most important developments of the 17th century, the re-emergence of atomism, was built in part upon chymical ideas and observations. Already in the late 13th century, the Latin alchemist known as Geber used a quasi-particulate matter theory to explain chemical properties. He explained, for example, gold's density and resistance to corrosion by theorizing that its 'tiniest parts' were tightly packed together leaving no space between them. Iron was more loosely packed, leaving spaces that rendered the metal lighter in weight and providing places for fire and corrosives to enter the metal and break it apart into rust. Later chymists continued to develop the idea of stable, minute particles, and to use it for explaining their observations. Mainstream Aristotelians often rejected such notions, for they claimed that substances lose their identity when combined. But practising chymists knew that they could often recover starting materials at the end of a sequence of transformations. For example, chymists knew that silver treated with acid 'disappears' into a clear, homogeneous liquid that passes freely through filter paper. When treated with salt, that liquid precipitates a heavy white powder, and that powder when mixed with charcoal and heated to red heat, gives the silver back again in its original weight. This well-known experiment indicated that the silver maintained its identity throughout, despite appearances and despite having been broken into invisibly small particles able to pass through the pores in paper. Chymical operations provided the best evidence for such 'atoms'.

Atomism and mechanism

The chymical tradition of particulate conceptions of matter cross-pollinated with a revival of ancient atomism. Ancient Greek atomism began with Leucippus and Democritus in the 5th century BC. They conceived of a material world composed of indivisible atoms moving in void space, their coming together and moving apart in ever-changing combinations gave rise to all the changes we see. Their conceptions largely died out in antiquity. Aristotle refuted them at length, and although Epicurus (341–270 BC) made atomism foundational for his moral philosophy, when Epicureanism fell out of favour because of its tendencies to atheism and hedonism (Epicurus intended neither consequence), atomism went out with the bathwater. A revival occurred only after the rediscovery of Lucretius' poem *On the Nature of Things*, a Roman popularization of Epicurus, in 1417. But Lucretius' emphasis on the link between atomism and atheism initially rendered his book unpalatable. Ironically, the rehabilitation of Epicurean atomism was due to a priest, Pierre Gassendi (1592–1655). Gassendi denied that atoms are eternal (only God is eternal) and that they move of their own accord (God set them moving), argued for the immateriality and immortality of the human soul, and then built a comprehensive world system using invisible particles and their motions as its fundamental explanatory principle. His system and others like it came to be called the 'mechanical philosophy'.

The mechanical philosophy holds that all sensible qualities and phenomena result from the size, shape, and motion of invisibly small pieces of matter – variously called atoms, corpuscles, or simply particles. Strict mechanical philosophers maintained that there is only one sort of 'stuff' out of which everything is made, and that only the differing shapes, sizes, and motions of the tiny particles of this single element provide the variety of substances and properties we perceive. Coherent with his disregard for qualities in favour of quantities, Galileo argued that most qualities, like hot and cold, colours, odours, and tastes do not actually exist, but are only the result of how minute particles affect our sense organs. For Galileo, and for later mechanical philosophers, the only real qualities – the *primary* qualities – were the size, shape, and mobility of particles. All other qualities were *secondary*, having existence only in the sensor, not the sensed. For the mechanist, vinegar seems sour only because its sharp and pointy particles prick the tongue. Apart from the tongue, 'sourness' has no meaning. A rose appears red only because of the way its particles modify reflected light and the way that modified light acts upon our eyes. The rose's pleasant smell results from an effluvium of particles the flower emits, that travel through the air into our nose where they strike the olfactory organ, producing motions which, when conveyed to the brain, are converted into a sensation of smell. This viewpoint fundamentally opposes Aristotelian ways of viewing the world, wherein sensible qualities have real existence in objects, and play a crucial role in explaining the object's nature and effects.

This system was mechanical in two senses. First, effects were caused only by mechanical contact – like a hammer on stone, or billiard balls colliding. There is no room for action-at-a-distance or powers of sympathy. Second, the world and objects in it – even plants and animals in the widely influential mechanical philosophy of Descartes – were conceptualized as *machines*. Mechanical philosophers compared the world to a complex clockwork – like the huge mechanical clocks of the period where hidden gears, weights, pulleys, and levers caused visible hands to turn, bells to ring, figurines to dance and bow, and mechanical roosters to crow, all in perfect order and regularity. The term 'machine of the world' (*machina mundi*) dates back to Lucretius and was used in the Middle Ages to express the complex regularity of the universe, but for those authors *machina* meant something more like frame or fabric, and expressed the interdependence of the various parts of creation. Mechanical philosophers however, gave the image the sense of an automaton, that is, something artificial but imitating the actions of a living thing mechanically. Mechanical perspectives reflected the increased technological prowess of the day, and shifted conceptualizations of the world away from living biological models towards lifeless machinery. This viewpoint led even to a reconceptualization of God Himself. Rather than a geometer, chymist, or architect, God was seen increasingly as a mechanic or watchmaker – a technician who designed and assembled the world machine. This image, which became particularly entrenched in late 17th-century England, forms the ultimate background to modern-day discussions of 'intelligent design'. In the early modern period, when theology and natural philosophy shaded seamlessly into one another, scientific and religious concepts grew and developed hand in hand, each one affecting and responding to the other.

As mechanical philosophers strove to apply their principles to all natural phenomena, one particular challenge was to explain the 'hidden qualities', sympathies, and actions-at-a-distance that had frustrated Aristotelians and formed the basis of natural magic. Mechanists' favoured solution was an appeal to invisible material effluvia – 'steams' of particles that carried effects from one body to another. For example, fire can heat an object at a distance because rapidly moving fire-particles emanate from the flame and strike the object. Other explanations required more inventive solutions. Descartes explained magnetic attraction by suggesting that magnets emitted a constant stream of screw-shaped particles. Iron, he postulated, contains screw-shaped pores, hence, the particles emitted by the magnet enter iron's pores and turn in them, thereby 'screwing' the iron closer to the magnet. Even the reflex action of turning away from a gory sight was explained on the basis of an efflux of sharp particles that wound the eyes.

Robert Boyle not only gave the mechanical philosophy its name, but joined it to chymistry in particular, recognizing chymistry's special ability to reveal the workings of the world. Boyle pursued all four major aspects of 17th-century chymistry: chrysopoeia, medicine, commerce, and natural philosophy. He sought avidly for the secret of making the Philosophers' Stone and tried to contact 'secret adepts' who could offer assistance. He claimed to have witnessed the Stone's use and tested the gold that he saw it produce from lead, and was responsible for having an English law forbidding transmutation repealed in 1689. He collected new chymical medicines, especially less expensive ones valuable for the relief of the poor (medical care and pharmaceuticals were overpriced then just as today). He also advocated the application of chymistry towards useful ends, for the improvement of trades, commerce, and manufacture. Perhaps most famously, he promoted chymistry as the best means for studying the world, and strove to elevate chymistry's status. Boyle explained that he devoted himself to chymistry, which his friends considered 'an empty and deceitful study', because it provided the best evidence for the particulate systems proposed by mechanical philosophers. As an example, he showed experimentally how saltpetre could yield both a fixed alkaline salt and a volatile acidic liquid, and how combining the two regenerated the saltpetre. The conclusion he drew was that compound substances could be taken apart into pieces and the pieces put back together to reform the original substance, just like parts of a machine. Although Boyle rejected much of Paracelsianism, such 'reintegrations' (as he called them) bear a striking resemblance to *spagyria*, and indeed, Boyle built his ideas upon a long foregoing tradition of both chrysopoeia and chemiatria.

The mechanical philosophy waned by the end of the 17th century. Boyle himself became less enthusiastic about it as he realized its overextension could lead to determinism, materialism, and atheism. If the world were only an array of colliding particles, there would remain no room for free will or divine providence. If God is a clockmaker, did He start the world running and then abandon it, or must He regularly readjust it as if He were less than a master mechanic? Chymists remained unimpressed by a strict mechanical philosophy – the vast array of properties they saw everyday did not seem explicable by the lean notions of a single kind of matter with differently shaped particles. Life processes were likewise far too complex for simple mechanics to explain beyond a certain point. Finally, Newton's forces of attraction, a kind of action-at-a-distance, were not reducible to mechanical explanation. The triumph of Newtonianism in fact meant the defeat of strict mechanism.

Evolving Aristotelianism

Aristotle and Aristotelianism have come in for quite a few knocks throughout this chapter. Indeed, one interpretation of the Scientific Revolution is that it was all about the rejection of a moribund Scholastic Aristotelianism. But this view fails to acknowledge the flexibility and continuing evolution of Scholasticism. While proponents of various 'new' philosophies of the 17th century routinely caricatured and criticized Aristotelianism with harsh rhetoric, other natural philosophers remained within that 'Aristotelian' framework and continued to update the system and to work productively. Neither in the late Middle Ages nor in the early modern period did being 'Aristotelian' or 'Scholastic' mean holding stubbornly to every claim made by Aristotle himself. Even Aristotle's own greatest student, Theophrastus, continued the Aristotelian tradition by disagreeing with his master on several points. In the Middle Ages, natural philosophers universally cited Aristotle, but often simply as a starting-point for their own explorations which frequently came to conclusions contrary to Aristotle's. By the Renaissance, there were many different and even conflicting Aristotelianisms.

Experimental and mathematical approaches to natural philosophy were not key parts of Aristotle's own work, but they increasingly became so for 17th-century Aristotelians. The Jesuits provide the clearest example of an explicit commitment to maintaining an Aristotelian natural philosophy, yet many, like Riccioli and Grimaldi, carried out extensive experiments relating to Galilean kinematics, and incorporated ideas and findings expressly contrary to Aristotle. Similarly, Niccolò Cabeo (1586–1650) rejected Gilbert's pro-Copernican interpretation of his magnetic experiments, but Cabeo's own experiments with the magnet were as extensive. By the end of the century, Jesuits had adopted many of the particulate and mechanical views expounded by Gassendi and Descartes, but within an 'Aristotelian' framework. For its proponents, Scholasticism remained a useful and flexible *method* of proceeding in the study of nature not necessarily a body of conclusions. While retaining a conservative stance towards the many innovations of the 17th century, they were nevertheless active participants and contributors to the Scientific Revolution.

What certainly did happen in the Scientific Revolution is that Aristotelianism acquired serious and radically different competitors, something it had not encountered in the late Middle Ages. Throughout the early modern period, new worldviews – the magnetical, chymical, mathematical, natural magical, mechanical, and others – emerged as challengers and plausible alternatives, while Scholasticism endeavoured to incorporate new material and ideas within an 'Aristotelian' framework. The continuing arguments between defenders of the various world systems resulted not only in a wealth of polemical pyrotechnics but also in a broad range of eclectic responses to the pressing challenge of establishing a new, and preferably comprehensive philosophy of nature. From our modern perspective, it is hard to imagine the broad diversity of viewpoints and approaches in regard to fundamental questions and methods that flourished in the early modern period, or the fertility and fervency with which an ever-increasing number of natural philosophers explored their world and devised systems – some small, some vast – to try to make sense of it all. This is one of the important ways in which the period of the 16th and 17th centuries was in fact 'revolutionary'.

Chapter 5
The microcosm and the living world

In addition to the world beyond the Moon and the world beneath the Moon, there was a third world that riveted the attention of early modern thinkers: the *microcosm* or 'little world' of the human body. Early modern physicians, anatomists, chymists, mechanists, and others focused on this living world that we inhabit. They explored its hidden structures, endeavoured to understand its functions, and hoped to find new ways of maintaining its health. The life that characterizes the human body naturally connects it to the rest of life on Earth – its flora and fauna. The catalogue of these living creatures exploded during the Scientific Revolution, thanks not only to voyages of exploration but also to the invention of the microscope, which revealed unimagined worlds of complexity in ordinary objects and new worlds of life within a drop of water.

Medicine

The human body was the first concern of the physician, and medicine had a high profile both socially and intellectually throughout the High Middle Ages and the early modern period. Alongside law and theology, medicine formed one of the three higher faculties of the university. The medical knowledge taught in 1500 was an accumulation of medieval Arabic and Latin experience and innovation built upon a core of ancient Greek and Roman teachings. Galen, Hippocrates, and Ibn Sīnā (or Avicenna,

c. 980–1037) stood as its chief authorities, and humoral theory formed its foundations. Humoral theory maintained that bodily health depended not only upon the proper functioning of the various organs, but also upon a balance, called *temperament*, among four 'humours', or fluids, found in the body: blood, phlegm, yellow bile, and black bile. These four humours corresponded with the four Aristotelian elements and shared their pairings of primary qualities (Figure 13).

The physician's role was to assist nature in re-establishing humoral balance by prescribing particular diets, daily regimens, and medicines. This predominantly Galenic medicine worked by 'contrary cures', that is, if a patient has (what we still Galenically call) a 'cold', resulting from excess phlegm (the cold and wet humour), then hot and dry foods and medicines should be administered to help restore balance. For a fever, cold and wet medicines are needed, cold baths, or perhaps bleeding to withdraw excess blood and its hot quality.

The many relationships held to exist between the superlunar world and the human body beautifully illustrates the connectedness of the early modern world. The macrocosm's influence on the microcosm was largely unquestioned, even if the details of this interaction were constantly debated. Thus astrology played a key role in both diagnosis and treatment; medicine, not prognostication, was probably astrology's chief application. Each bodily organ corresponded with a zodiacal sign and was particularly susceptible to influences from the planet that resembled it in qualities (Figure 14).

The brain, for example, a cold and wet organ, is influenced most by the Moon, a cold and wet planet. (Hence, someone with disordered brains is still today called a lunatic – from *luna*, Latin for Moon – or more colloquially, 'moony'.) Knowing the planetary positions at the onset of an illness could therefore assist in diagnosis by helping the physician understand prevailing

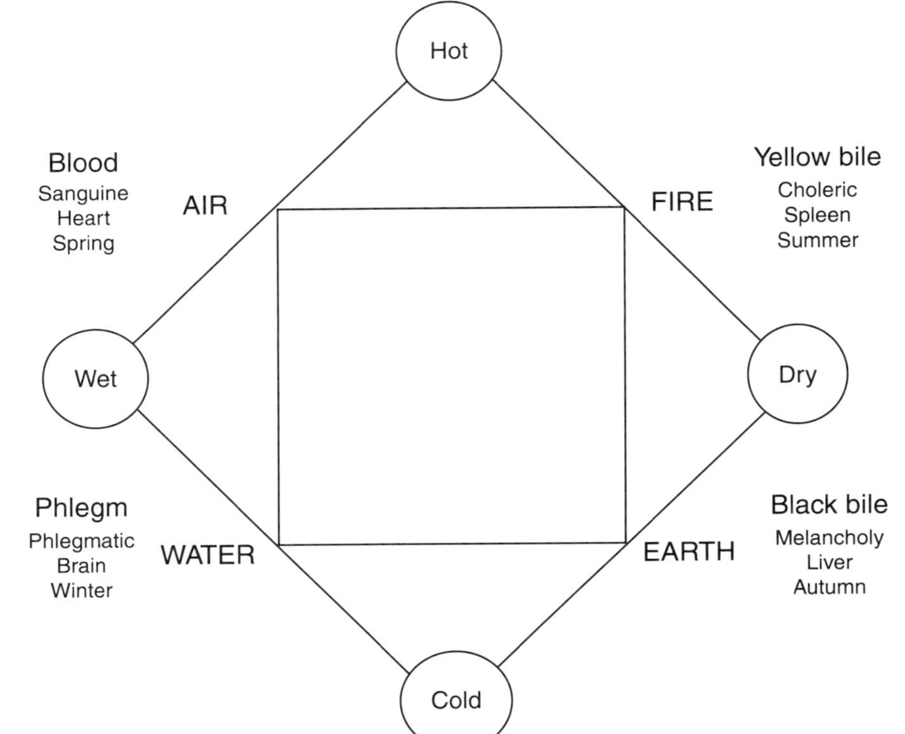

13. A 'square of the elements' showing their qualities and their relationships with the four humours, four bodily complexions, and four seasons

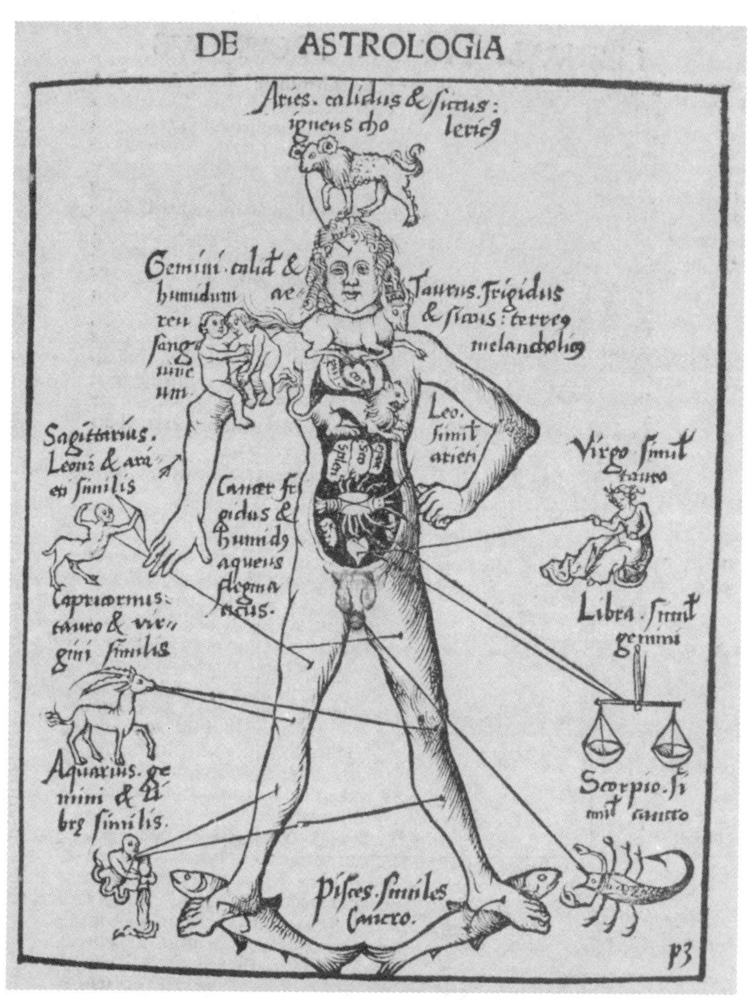

14. A chart of the organs and their zodiacal correspondences. From the early modern encyclopaedia compiled by Gregor Reisch, entitled *Margarita philosophica* (Freiburg, 1503)

environmental influences or localize potentially affected parts of the body. Furthermore, each person was held to have a unique ratio of humours – called his *complexion* – imprinted at birth by the then-prevailing planetary influences; this means that every patient must be restored to his own particular complexion.

One size does not fit all in early modern medicine. Medical treatments had to be tailored to each patient; the same pill could not be used on everyone, and a particular diet and regimen had to be followed in parallel with treatment. A physician might therefore examine the patient's natal chart to gain insight about the patient's complexion. Astrological calculations could also assist in timing medical treatments, according to the Hippocratic idea of 'critical days', namely, that during the progress of an illness there are points of 'crisis' that must be successfully overcome for the patient to recover. Diagnosis also relied on the examination of urine – portable reference charts provided tables of the colour, smell, consistency, or even taste of patients' urine and the relation of these indicators to various ailments. The same is true of the rate, rhythm, and strength of the pulse.

Methods of medical treatment, at least among licensed physicians, did not change dramatically during the Scientific Revolution. Despite a slow evolution in response to new ideas and discoveries, the core of Galenic and Hippocratic medicine continued well into the 18th century (although astrological diagnoses began to wane in the 17th). This endurance reflects both the stability of medical school curricula and the guild or licensing structure of medicine that promoted conservatism. Innovations thus often came from outside the body of licensed physicians. The strict licensing of physicians was, however, possible only in large urban centres. In most places, a variety of healers with little or no formal medical education attended to people's health and far outnumbered licensed physicians. Virtually every householder kept a list of home remedies for family and neighbours. Apothecaries made both simple and prepared medicines easily available such that virtually anyone could compound even exotic (and sometimes dangerous) medicines. Surgeries were carried out by barber-surgeons, a group with lower status and less formal training than physicians. 'Empirics', unlicensed physicians offering a variety of medicaments and treatments, found their best trade in the cities despite frequent attempts to ban them from London, Paris, and other major centres. In dramatic contrast to modern medical practice, some treatments were done contractually, that is, the physician's remuneration was dependent upon his success.

Radically new medical approaches such as Paracelsianism and, in the 17th century, Helmontianism were taken up more avidly by unlicensed practitioners, often in direct challenge to the medical establishment. Nevertheless, new chymical approaches to medicine made slow but steady ingress into official pharmacopiae and the practices of professional institutions like the Royal College of Physicians of London, established in 1518. In France, the conservative Galenic faculty of medicine in Paris and the pro-Paracelsian faculty at Montpellier waged a decades-long battle over the risks and rewards of chymical medicines. This conflict also reflected fault lines running between the royal, centralized, and predominantly Catholic Parisians, and the provincial, mostly Protestant Montpellians. Their most fervent debate, over the medical use of antimony – a toxic mineral – came to an end only after 1658 when Louis XIV, having fallen ill during a military campaign and not responding to traditional treatments by the royal physicians, was cured by a vomit induced by a dose of antimony in wine administered by a local physician. The Parisian medical faculty thereafter had little recourse but to vote to legalize the use of this Paracelsian *vin émetique*.

Anatomy

Anatomy witnessed significant development in the early modern period. Although Galen stressed the importance of anatomy in antiquity, Romans considered the violation of dead bodies by dissection socially and morally unacceptable, and thus Galen dissected apes and dogs and transferred his findings by analogy to human beings. (Nevertheless, he undoubtedly saw exposed human innards from time to time during his position as a physician to gladiators.) Only in Egypt were human dissections carried out in antiquity, probably because opening the body and removing its organs was already familiar there due to the practice of mummification. During the late Middle Ages, however, human dissection became standard in Italian medical schools such as Padua and Bologna. By about 1300, medical students were required to observe a human dissection as part of their training. There is no basis whatsoever to the 19th-century myth that the Catholic Church prohibited human dissection. Human dissection was hampered mostly by a shortage of corpses. Since respectable people would not permit their bodies or those of their kin to be displayed and cut up before an audience, dissections were dependent upon the availability of corpses from executed criminals, often foreigners.

Interest in human anatomy increased greatly in the early 16th century, particularly in Italy, culminating in Andreas Vesalius' (1514–64) monumental work *On the Structure of the Human Body*, published in 1543, the same year as Copernicus's *On the Revolutions*. Born in Flanders, Vesalius trained at Padua and became lecturer of surgery there the day after receiving his MD. Assisted by a judge who timed executions conveniently (without refrigeration or preservatives, corpses had to be dissected immediately), Vesalius performed many careful dissections, noting the errors of Galen and other authors, and grouping parts of the human body in new ways, no longer just functionally but structurally as well. Drawing upon the skills of artists from Titian's workshop, Vesalius supervised the production of detailed anatomical drawings, and these formed a main feature of his book, whose text explained each illustration and anatomical feature in great detail. Producing so richly illustrated a book would have been impossible without the printing press. Still, the lavish volume was expensive, spurring Vesalius to produce a cheaper version for students, through which his ideas, discoveries, and organizing principles gained wide circulation. Increased interest in anatomy led to the construction of anatomy theatres, first in Padua (1594), then in Leiden (1596), Bologna (1637), and elsewhere. Although intended for teaching medical students, these theatres, especially those in Northern Europe, attracted large audiences of interested (or fashionable) onlookers from the wider public as well.

Dissections were not restricted to either human corpses or medical schools. With the rise of 17th-century scientific societies, animal dissections became a major part of their activities. In the 1670s and 1680s, the young Parisian Royal Academy of Sciences received the bodies of exotic animals that had died in Louis XIV's menagerie, including an ostrich, lion, chameleon, gazelle, beaver, and camel. While dissecting the last of these, the head of the Academy, Claude Perrault (1613–88), nicked himself with the scalpel and died from the resulting infection. In the 1650s and 1660s at Oxford, and then at the Royal Society in London, several workers dissected not only dead but living animals, especially dogs, in experiments too gruesome for the modern reader to stomach (Boyle himself was disturbed by them). These experiments endeavoured to learn the actual workings of nerves, tendons, lungs, veins, and arteries. Often, they included the injection of various fluids to observe their movement through the body and their physiological effects, as well as blood transfusions, sometimes from one species into another, including attempts to cure sick human beings with blood transfused directly from healthy sheep.

This interest in blood and the movement of bodily fluids stemmed in part from William Harvey's (1578–1657) arguments for the circulation of the blood published in 1628. According to Galen, the venous and arterial systems are separate units. The liver continuously produces dark venous blood that the veins distribute through the body as nutriment. A portion of this blood is drawn into the heart, where it passes through pores in the tissue, or septum, dividing the right and left ventricles. There, air drawn from the lungs via the pulmonary artery converts it into bright arterial blood, which then nourishes the body through the arterial system. No blood ever returns to the heart. The 16th-century anatomists, however, found problems with Galen's system. They questioned the existence of pores in the septum, and found that the pulmonary artery was full of blood, not air. The latter observation led to the proposal of the 'lesser circulation': venous blood passes from the heart through the lungs, then returns to the heart before flowing out into the body. At the University of Padua, Harvey studied with the greatest anatomists of the day, notably Girolamo Fabrizio d'Acquapendente (1537–1619), who had described 'valves' he found in the veins. Harvey later remarked that this discovery led him to consider a wider circulation of the blood.

Harvey noted that the volume of blood pumped by the heart would exhaust the body's supply within moments unless it was somehow recirculated. Using ligatures to stop the flow of blood selectively, he experimentally deduced the 'greater circulation', namely, that the heart pumps blood circularly through connected arterial and venous systems. Harvey found the blood's circular motion satisfying, since it meant that the microcosm mimicked the macrocosmic heavens whose natural circular motion Aristotle considered the most perfect. Indeed, Harvey maintained Aristotelian approaches and methods, and focused attention on the heart and blood partly because of the central role Aristotle had given them – another example of Aristotle's continuing importance in the Scientific Revolution. Harvey was unable, however, to detect the tiny capillaries that connect arteries to veins. These structures were first seen only four years after Harvey's death by Marcello Malpighi (1628–94), who observed the movement of blood through minute vessels linking the pulmonary vein to the pulmonary artery in the transparent lung tissues of frogs; he extrapolated that similar vessels connected arteries to veins throughout the body. To make this observation, Malpighi used a relatively new invention: the microscope.

Microscopy, mechanism, and generation

The origins of the microscope in the early 17th century are obscure, but like its sister the telescope, it revealed a new world and provoked new ideas. Galileo used a device similar to his telescope to magnify small objects, but the first drawings made using the microscope appear in studies of bees carried out in 1625 by Francesco Stelluti and Federico Cesi and dedicated to Pope Urban VIII, whose Barberini family used the bee as an emblem. In the 1660s, Robert Hooke built an improved microscope to examine everything from tiny insects like lice, to frost crystals, and the fine structure of cork, which he found divided into chambers he called 'cells' after their resemblance to monastic living quarters. Antoni van Leeuwenhoek (1632–1723), a draper and surveyor in Delft, devised the simplest and most powerful magnifiers. He built more than five hundred microscopes using a tiny spherical glass bead as their single lens, and published more microscopical observations than any other author. He subjected an incredible array of objects to his microscopes, observing 'worms' in human and animal semen, corpuscles in blood (and their movement through capillaries in the tail of a young eel), bacteria in dental plaque, and swarming 'animalcules' in pond water and infusions of vegetable matter. His discovery of spermatozoa fed into a lively debate over the nature of animal and plant generation. Leeuwenhoek himself supported *preformationism*, the idea that a tiny version of new offspring was contained within each spermatozoon, or, according to some of his contemporaries, within each egg. The opposite view, *epigenesis*, held that embryonic structure was produced *de novo* and in successive stages during gestation. Preformationism appealed especially to mechanical philosophers because it reduced generation to a simple matter of mechanical growth – a tiny organism simply got bigger by assimilating new matter. As such it abandoned the immaterial vital forces most epigenesists considered necessary for crafting amorphous material – semen and/or menstrual blood or the fluid of an egg – into an organized and differentiated embryo. Harvey, an epigenesist, by opening chicken eggs at various stages of their development, observed that blood formed first, which he took as evidence that it was the seat of life and of a vital soul that guided the formation of the offspring. Preformationism however provoked the question of where and when the tiny form of the new organism actually began. A few suggested that all future generations were contained, one inside the next, within the first of a species created by God.

The microscope's revelation of seemingly mechanical structures in living bodies excited mechanists in particular, and accordingly most microscopists of the late 17th century were mechanists. They embraced Harvey's circulation of the blood in part because it characterized the heart as a pump – a mechanical device, although Harvey was far from a mechanist himself – and they strove to reduce complex living systems to mechanical principles. In Florence, Giovanni Alfonso Borelli (1608–79), for example, analysed animal motion in terms of simple machines – conceptualizing bones, tendons, and muscles as levers, fulcrums, and ropes, and bodily fluids and vessels as hydraulics and plumbing, thus launching what has come to be called biomechanics. In London, Nehemiah Grew (1641–1712) explored the hidden anatomical structures of plants, helping to establish plant physiology. Some mechanists even hoped that improved microscopes would allow the direct observation of atoms, their shapes, and their motions, exposing to direct observation the fundamental explanatory principles of the mechanical philosophy.

Microscopical observations, like all others, were open to conflicting interpretations. While the discovery of spermatozoa could be interpreted to favour preformationism, the contemporaneous discovery that countless living creatures appeared on their own in stale water strongly favoured established notions of spontaneous generation – that living creatures could emerge from non-living material – which in turn favoured the epigenic idea that living structures emerge from originally amorphous matter. For centuries previously, most natural philosophers had assumed that simple life forms appeared spontaneously under certain circumstances – a rotting bull carcass generated bees, mud generated worms, putrefying flesh generated maggots. In a series of famous experiments in the 1660s carried out at the Medici court,

Francesco Redi (1626–97) left samples of meat out to rot, some covered with a mesh or cloth and others in the open air. Those in the open air produced maggots, while none appeared when access by flies was prevented. As in most cases of experiments seen retrospectively as 'definitive', Redi's experiments did not immediately stamp out belief in spontaneous generation since other explanations of the results could be (and were) offered, and Redi himself allowed that some insects – like the oak gall wasp – might be produced directly from plant matter. Although moderns routinely scoff at belief in spontaneous generation, it is worth pointing out that any modern scientist who does not believe in a special creation of the first life form by God's miraculous intervention must consequently believe in the spontaneous generation of life from non-living matter.

Neither the microscope nor the mechanical view of living systems lived up to expectations. The limits of magnification and resolution, given the material and optical systems available, were soon reached. Microscopical investigation had revealed such enormous complexity in living systems that mechanistic explanations seemed increasingly inadequate to account for either their formation or their functioning. Yet even while mechanical approaches enjoyed their greatest popularity, more vitalistic models also flourished. In fact, the divide between non-living and living was not at all clear-cut in the 17th century, and many thinkers hybridized mechanical and vitalistic systems. For example, few mechanists were so rigid that they denied the existence of an animating soul in living systems. Such a soul need not be like the immaterial, immortal human soul of Christian theology, but rather was considered to exist in various forms or levels in various entities (for modern readers perhaps the term 'vital spirit' better expresses the concept). These notions date back to Aristotle, who had proposed three levels of soul: a *vegetative* soul in plants responsible for overseeing growth and the assimilation of nutrition; in animals, a further *sensitive* soul to govern sensation and movement; and in human beings, in addition to the vegetative and sensitive soul, a *rational* soul to govern thought and reason. For many, while mechanical principles could explain particular bodily functions and structures, the organization and maintenance of the organism as a whole – not to mention consciousness and awareness – were functions of soul.

Helmontianism

Perhaps the most comprehensive new system of medicine to emerge in the 17th century was that of the Flemish nobleman, physician, chymist, and natural philosopher Joan Baptista van Helmont (1579–1644). Van Helmont combined chymistry, medicine, theology, experiment, and practical experience into a cohesive and highly influential system. His autobiographical statements express a dissatisfaction with traditional learning and a desire to pursue new knowledge that is typical of Scientific Revolution-era thinkers. He recounts how he attended the University of Louvain, but refused his degree because he felt he had learned nothing. He then studied with the Jesuits and felt no better off. Then he obtained an MD, but finding the foundations of medicine 'rotten', he turned to Paracelsianism, only to reject much of that as well. Thus van Helmont endeavoured to start afresh, calling himself a 'philosopher by fire', meaning that his training came not from traditional learning but from experiments in chymical furnaces. Indeed, van Helmont was an extraordinary observationalist, describing the origin, symptoms, and progress of several maladies that were not otherwise recognized until centuries later.

Van Helmont rejected the four Aristotelian elements and the Paracelsian *tria prima*, claiming instead that water was the single underlying element of everything. Not only did this idea harken back to the oldest-known Greek philosopher Thales, but more importantly (for van Helmont) also to Genesis 1:2 where the Spirit of God brings forth the world by 'brooding [like a hen] upon the *waters*'. The Belgian philosopher sought experimental confirmation of this idea, most famously by planting a five-pound willow sapling in two hundred pounds of soil, and watering it for five years. At the end of that time, the tree weighed 164 pounds but the soil had lost scarcely any weight; therefore, van Helmont concluded, the entire composition of the tree must have been produced from water alone. According to van Helmont, *semina* (seeds) implanted in the world at creation have the power to transform water into all substances. These *semina* are not physical seeds like a bean, but immaterial organizing principles, like the invisible vital principle that organizes the fluid of an egg-yolk into a chick. Fire and putrefaction destroy the *semina* and their organizational power, thus turning substances into air-like substances van Helmont called 'Gas' (from the word *chaos*, and the direct source of our word for the third state of matter). Thus burning charcoal and fermenting beer release a choking *Gas sylvestris*, and burning sulphur a stinking *Gas sulphuris*. In the cold parts of the atmosphere, this *Gas* finishes converting back into primordial water and falls as rain, thus closing the cycle of water's successive transformations in van Helmont's economy of nature.

Similar to, but more sophisticated than Paracelsus, van Helmont held that bodily processes were fundamentally chemical. He recognized the acidity of gastric juice responsible for digestion, and performed analyses of bodily fluids – especially of urine to find the cause and cure of one of the 17th century's most painful maladies, kidney and bladder stones. Yet chemical processes could not suffice on their own to explain life processes; they had to be directed by a quasi-spiritual entity lodged in the body and called the *archeus*. For van Helmont, the *archeus* regulates and governs bodily functions. Sickness results from a weakened *archeus*, unable to perform its duties; medical treatment must therefore work to strengthen the *archeus*. Accordingly, van Helmont rejected Galenic notions of complexion, the four humours, and methods of healing. Diseases like the plague, he said, are not due to humoral imbalance but to external 'seeds' of disease invading the body and transforming it. A strong *archeus* can dispel these seeds, but a weak one needs help. (Note that in both Galenic and Helmontian medicine, the physician's role is always to *assist* natural processes, never to divert them or to assert control over the body.) Van Helmont also emphasized the role of the patient's mental and emotional state, and claimed that the power of imagination can cause physical changes in the body. Helmontian ideas deeply influenced many physicians, physiologists, and chymists.

Mechanist and vitalist conceptions of living systems were not irreconcilable but rather two ends of a continuum; many physicians and natural philosophers embraced intermediate positions. Like his contemporary van Helmont, Gassendi invoked seeds as powerful principles able to organize matter into new forms. But while van Helmont's seeds were immaterial, Gassendi's were special combinations of physical atoms (divinely organized) that acted mechanically on matter. Indeed, mechanist and vitalist speculations produced hybrid medical systems in the 18th century, such as that of Georg Ernst Stahl (1659–1734) which emphasized the mechanical nature of chemical transformations but the need for vital powers to organize and govern living systems. Herman Boerhaave (1668–1738), perhaps the most influential voice in 18th-century medicine, especially in pedagogy, drew together diverse strands of 17th-century natural philosophy. As professor of medicine and chemistry at Leiden University's medical school, Boerhaave strongly advocated both Hippocratic methods of healing (emphasizing environment and patient individuality) and the importance of chemistry for medical education. His approach to medicine and the body combined aspects of Boyle's mechanical philosophy, Newton's physics, and van Helmont's 'seeds'. Boerhaave's reforms of medical education were adopted throughout much of Europe (hence he was sometimes called the 'Teacher of Europe'), and proved foundational for significant changes that would occur in 18th-century medicine.

Plants and animals

The study of flora and fauna – what we would call botany and zoology – expanded enormously in the 16th and 17th centuries. The traditional textual location for such material was an encyclopedia tradition, stemming from the massive *Natural History* compiled by Pliny the Elder (23–79 AD) in his attempt to collect and popularize Greek learning for a general Roman audience. Encyclopedic accounts of plants and animals filled medieval herbals and bestiaries, and this format continued into the Scientific Revolution. One of the most famous is the five-volume *History of Animals* by Conrad Gessner (1516–65) with its hundreds of woodcuts. Many such volumes would, however, appear strange to modern readers, for they blend naturalistic and descriptive details about various species with a mass of literary, etymological, biblical, moral, mythological, and metaphorical meanings that had accumulated around each animal or plant since antiquity. No account of the peacock would be complete without mention of its pride, of the serpent without its deceitful role in Adam's Fall, of the plantain (a common plant that grows in footpaths) without reference to how it signifies the well-trodden way of Christ. Plants and animals are not presented as isolated species, but rather within rich networks of meaning and allusion. They are both natural objects and emblems dependent upon the vision of a world of layered meanings, a world simultaneously literal *and* metaphorical, a world full of symbolic messages to be read. As a result, even fabulous animals such as the unicorn, dragon, and various monsters are described alongside well-known creatures, not necessarily because the authors believed they roamed the Earth, but more because whether or not they existed in the physical world, they nevertheless carried meaning thanks to their existence in the literary world. While modern readers might consider such texts 'quaint', credulous, or encumbered with 'non-scientific trivia', their original audience would probably consider modern botanical or zoological descriptive texts sterile and oddly disengaged from humanity.

Two developments of the early modern period diverted this emblematic tradition into other directions. The needs of medicine to identify herbal remedies was the first. As humanist scholars continued to revive, edit, and publish Greek medical texts, it became increasingly necessary to identify the medicinal plants these texts mentioned and to help locate them in the wild. Hence there was a demand for new herbals that bridged the gap between ancient texts and what grew in 16th-century fields. To accomplish this task, new herbals not only linked common names with ancient Greek ones, but provided accurate, naturalistic illustrations of them. Just as Vesalius collaborated with artists from Titian's workshop, so too a new generation of 16th-century botanists worked with artists to produce herbals with extensive illustrations drawn from life, such as Otto Brunfels' *Living Images of Plants* (1530–6) and Leonhart Fuchs' *History of Plants* (1542). The second development was the expansion of European horizons. On the narrowest level, ancient authorities like Dioscorides described mostly Mediterranean plants and did not recognize Northern European species, hence it became necessary to provide accounts of plants that did not have a Classical pedigree. The same problem, but on a much larger scale, existed in terms of the countless plants and animals encountered for the first time in voyages outside of Europe, especially in the Americas. Food plants like potatoes, corn, and tomatoes, medicinal plants like 'Jesuit's bark' (cinchona, the source of quinine, a cure for malaria), and new animals like the opossum, jaguar, and armadillo greatly increased the catalogue of flora and fauna known to Europeans. These new arrivals had no accumulated networks of correspondence and emblematics, and so could not be fit into the traditional format of herbals and bestiaries. In Spain, where most reports from the New World first arrived, those charged by the king with organizing the information were forced to give up established encyclopaedic methods based on Classical models like Pliny not only because new findings rendered old categories obsolete, but also because the unrelenting flow of new information made it impossible to organize the knowledge comprehensively.

Spaniards in the New World, often members of religious orders, struggled to chronicle native plants, animals, and medical practices, sometimes collaborating with indigenous scholars to produce illustrated texts. José de Acosta (1539–1600), sometimes called the 'Pliny of the New World', was a Jesuit in Peru who, besides founding five colleges, wrote a natural history of Latin America that was widely published, translated, and referenced in Europe. In 1570, King Philip II sent his physician Francisco Hernández on an expedition specifically to seek out New World medicinal plants. Hernández spent seven years, mostly in Mexico, cataloguing plants and inquiring about their properties from indigenous healers, while a team of native artists produced the illustrations for a six-volume *Plants and Animals of New Spain* (it describes about 3,000 plants and dozens of animals). Frustrated by the impossibility of inserting new plants into Classical classification schemes, Hernández even adopted native names to create a new botanical taxonomy. Meanwhile, the Franciscan friar Bernardino de Sahagún (1499–1590), working with Aztec assistants and informants at the Colegio de Santa Cruz at Tlatelolco in Mexico, produced the *General History of Things in New Spain*, a lengthy work in both Spanish and Nahuatl describing Aztec culture, customs, society, and language. At home in Spain, the physician Nicolás Monardes (1493–1588) compiled a *Medicinal History of Things Brought Back from Our West Indies* that described dozens of New World species. Portuguese scholars such as Garcia de Orta (1501–68) and Cristóvao da Costa (1515–94) likewise reported on their findings of medicinal plants and new animals in India and elsewhere in South and East Asia.

The search for new medicines drove the study of new plants, and consequently the establishment of botanical gardens, usually in the context of medical schools. Medicinal gardens had been a part of monasteries throughout the Middle Ages, and new botanical gardens built upon this foundation and expanded it for pedagogical and research purposes. The first botanical gardens opened in Italy at the universities of Pisa and Padua in the 1540s and Bologna in 1568, along with the establishment of professorial chairs in medical botany.

Other centres of medical instruction followed suit – Valencia (1567), Leiden (1577), Leipzig (1579), Paris (1597), Montpellier (1598), Oxford (1621), to name a few. These gardens were laid out in strict order, with species grouped by therapeutic property, morphology, or geographical origin. Seeds, roots, cuttings, and bulbs were sought after, traded, and exchanged, thereby expanding the range of plants available in gardens across Europe. The interest in unusual plant cultivation and hybridization spread to private individuals, leading to the celebrated 17th-century 'tulipomania' in the Netherlands, where newly made bourgeois fortunes were drained to acquire rare hybrids, and artists preserved exotic flowers in still-lifes.

A widespread interest in the exotic and rare expressed itself in collecting natural historical specimens of all kinds in 'cabinets of curiosities' (Figure 15). While these collections were in one sense

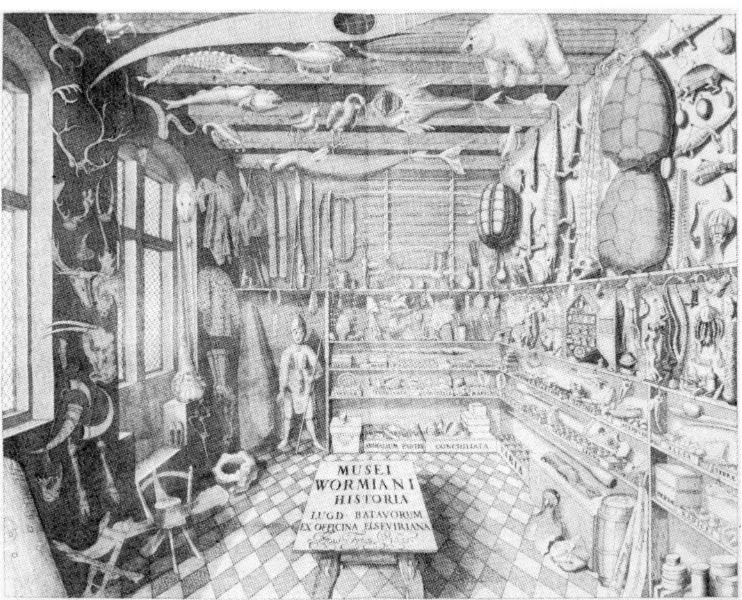

15. The cabinet of curiosities of Ole Worm from *Museum Wormianum (The Wormian Museum, or, A History of the Rarer Things both Natural and Artificial, Domestic and Exotic, which the author collected in his house in Copenhagen)* (Leiden, 1655)

the forerunners of museums, they also functioned to display the power, wealth, connections, and interests of their collectors and to invoke wonder at the marvels of nature and art. Princes and noblemen as well as scholars amassed collections that included both *naturalia* – botanical, zoological, and mineralogical specimens – and *artificialia* – mechanical contrivances, stunning works of art and craft, and ethnographic and archaeological objects. Ulisse Aldrovandi (1522–1605) compiled one of the earliest such collections (part of which still survives in Bologna), and a guided tour by Athanasius Kircher of his museum at the Collegio Romano was a 'must see' for 17th-century visitors to the Eternal City. The physical arrangements of the objects within the space of the cabinet emphasized the connections between objects – often ones that we would not consider. Thus these cabinets became microcosms of another sort, displaying and emblematizing the diverse, the marvellous, and the exotic of the linked worlds of man and nature all compressed into a single room.

Chapter 6
Building a world of science

Science is more than the study and accumulation of knowledge about the natural world. From the late Middle Ages down to our own day, scientific knowledge has been used increasingly to change that world, to give human beings greater power over it, and to create the new worlds in which we now live so much of our lives, seemingly ever more separated from the natural world. More and more people today become so surrounded by the world of artifice constructed by technology that they notice their dependence upon it only when it fails, and then find themselves as helpless as a medieval farmer when the rain does not fall on his crops. Thus moderns often react with consternation when the natural world reasserts itself by intruding inconveniently upon this artificial world – when meteorites or solar flares knock out satellite communications, lightning strikes cut off electrical power, or volcanic eruptions shut down airline traffic. The proliferation of technology has changed the daily world of human beings more radically than anything else in the last few centuries. That explosion of technology simultaneously depends upon and encourages scientific inquiry. The 16th and 17th centuries witnessed a special turn towards applying scientific study and knowledge to address contemporaneous problems and needs.

The world of artifice

In Renaissance Italy, ambitious new engineering projects transformed landscape and cityscape. Canals and waterworks claimed new land and provided drinking water and transportation routes. Filippo Brunelleschi's (1377–1446) immense double-shelled dome for the cathedral, with its innovative construction techniques set a new skyline for Florence. New urban design fulfilled the humanist emphasis on civic life and proclaimed the wisdom and power of ruling princes while new fortifications protected their interests. As is often the case, one new technology drove the development of others. The technological transformation of warfare in the 15th century by the increasing use of gunpowder and production of portable bronze cannons rendered medieval fortifications obsolete – their soaring battlements provided excellent targets for artillery. Thus a new system of fortification had to be developed. New designs for fortification drew upon geometrical principles and became standard parts of a nobleman's education. Pressing practical concerns (and princely ambitions) produced, first in 16th-century Italy, then elsewhere, a class of learned engineers and architects who, following the lead of the ancient models Archimedes and Vitruvius, increasingly turned to mathematical principles and analysis to solve practical problems. Falling between artisans who relied on accumulated manual experience and scholars removed from practical affairs, this emergent class provided a crucial background for the increasing deployment of mathematics to investigations of the world, an essential feature of the Scientific Revolution. Leonardo da Vinci (1452–1519) is one early example of this 'intermediate' group, as is the military engineer Tartaglia in the mid-16th century. At the end of the century Galileo drew inspiration and borrowed methods from the learned engineers.

Both practicality and the humanist desire to emulate the ancients inspired renovations of the city of Rome. Papally sponsored projects explored and rebuilt ancient acqueducts and sewers.

The dilapidated 4th-century St Peter's was pulled down to build the immense new basilica that stands there today, and provoked one of the most spectacular engineering feats of the 16th century: the moving of the Vatican obelisk. A single stone the height of a six-storey building and weighing over 360 tons, the obelisk had been erected by the Romans in the 1st century. In 1585, as the new St Peter's encroached upon the obelisk, Pope Sixtus V issued a call for proposals to move the ancient Egyptian stone to a new location – the first time an obelisk had been moved in 1,500 years. The engineer Domenico Fontana (1543–1607) won the commission. Using the combined power of 75 horses and 900 men operating 40 windlasses, five 50-foot-long levers, and eight miles of rope, Fontana successfully lifted the monolith – encased in iron armatures – straight up off its base on 30 April 1586. The operation was considered so important that the Pope allowed part of the newly completed basilica to be torn down in order to allow optimal operation of the levers and windlasses. Fontana then lowered the obelisk onto a carriage (Figure 16), transported it along a causeway, and re-erected it where it stands today, the focal point of St Peter's Square.

Renaissance achievements, and the economic and military engines that supported them, required materials. Accordingly, the period 1460 to 1550 witnessed a mining boom, particularly in central Europe where mineral resources were richest. Medieval mining had been largely a small-scale operation that exploited surface deposits. But the demands of early modern Europe – iron and copper for arms and artillery, silver and gold for coinage – drove more organized, larger-scale mining and the development of better smelting and refining techniques. Deeper shafts and increased scales required more mechanization – water wheels to drive bellows and rock-crushing equipment, pumps to drain mines and ventilate shafts – as well as greater organization of labour. Perhaps the most famous writer on mining, Georgius Agricola (1494–1555), a German humanist and educator, endeavoured to organize and promote mining knowledge. His massive and richly illustrated

16. Moving the Vatican obelisk, from Domenico Fontana, *Della trasportazione dell'obelisco vaticano* (Rome, 1590)

Latin treatise *On Metallic Things* endeavoured to ennoble an otherwise dirty enterprise by linking German mining practices with Classical literature and creating a Latin vocabulary for metallurgy. The landscape of felled trees, smoke, and streams of runoff shown incidentally in Agricola's illustrations underlines how such technological growth came with increasing costs to the environment. Probably more useful to actual practitioners were the German-language books of Lazar Ercker (c. 1530–94), an overseer of mining operations. His books are filled with practical experience about treating ores, assaying metals, and preparing chemical products like acids and salts including saltpetre, the crucial ingredient in gunpowder. By the mid-16th century, the boom was over – ended as much by the depletion of European mines as by the flood of metals from the New World that depressed metal prices, making the working of European mines less profitable.

The promise of the New World spurred developments in cartography and navigation. Late medieval navigational charts, or portolans, indicated only coastlines overlaid with rosettes of compass-headings from particular points. These charts were useful for relatively short journeys in the Mediterranean or along coastlines, but not for providing a geographical perspective or for journeying across oceans. Ptolemy's 2nd-century *Geography*, rediscovered in the fifteenth century, described using a grid of east–west and north–south lines (latitude and longitude, respectively) for mapping. Late 15th-century maps – such as Waldseemüller's – adopted this method, employing curved latitude lines and longitude lines that converged towards the poles. The Flemish cartographer Gerhardus Mercator (1512–94) popularized the now more familiar Mercator projection, where parallel longitude lines intersect straight latitude lines at right angles. Although it distorts land masses at high latitudes, this method of projecting the spherical Earth on a flat map was easier for navigation (at least at low latitudes) and was favoured by Spanish cosmographers and navigators.

The compass and quadrant – instruments to determine heading and latitude, respectively – had been used for navigation since the Middle Ages, but there existed no reliable method to determine longitude. This inability did not present a serious problem while vessels stayed in European waters or within sight of land. But crossing oceans was a perilous venture without accurate longitude measurements. Since locating a place requires both latitude and longitude, the lack of longitude presented so serious a problem for cartographers and navigators that finding a method to determine it became the most urgent technological problem of the period. Competing seafaring states – Spain, the Netherlands, France, and England – offered rich prizes for anyone who could devise a reliable method.

Time-telling is the key to longitude. Every hour of difference in local time between two places translates into fifteen degrees of longitude in separation (hence a modern 'time zone' is roughly fifteen degrees wide). But how to know the time at two distant locales simultaneously? One could take along a clock set at the ship's place of origin, and compare its reading with the time at the ship's location as determined by observations of Sun or stars. Unfortunately, early modern clocks were barely reliable to twenty minutes a day. Galileo's observation that pendula beat out a constant tempo regardless of the amplitude of their swing suggested a new regulator for time-keeping. He began designing a pendulum-regulated clock while under house arrest, but never built it. It was Christiaan Huygens in the Netherlands who produced the first workable pendulum clock in 1656, resulting in a huge leap in reliability – at least for land-based clocks. On a rocking ship, pendulum clocks did not run accurately. Thereafter, Huygens and Robert Hooke experimented independently with spring-powered clocks, but these too proved insufficiently accurate aboard ships. Still, Hooke's study of springs led to his enunciation of the relationship between the extension and the force of a spring, known today as 'Hooke's Law', just as Huygens's work led to refinements of the laws of simple harmonic motion. (The longitude problem itself was solved only in the 18th century using innovative chronometers devised by the English instrument-maker John Harrison that could keep accurate time even at sea.)

The alternative to a manmade clock was a celestial one – some astronomical event whose time of occurrence at a reference site could be calculated and then compared with the local time of the event at the observer's site. Spanish cosmographers of the 16th century successfully used coordinated observations of lunar eclipses to determine the longitude of settlements in the Spanish empire, but lunar eclipses are too rare for navigation. Jupiter's four moons, however, undergo more frequent eclipses – the innermost satellite Io has an eclipse every forty-two hours – and Galileo proposed using them as time-keepers. The astronomer Gian Domenico Cassini (1625–1712) explored this idea most fully, and in the 1660s compiled timetables of these eclipses. But once again, while this system worked on land – it was used successfully for correcting terrestrial maps – it proved impractical to observe the eclipses telescopically from a moving ship. Nevertheless, while testing the idea, observers noticed that some eclipses occurred several minutes later than predicted. Realizing that this discrepancy was greatest when Jupiter was farthest from Earth, the Danish natural philosopher Ole Roemer (1644–1710) proposed in 1676 that light has a finite speed – the eclipse's apparent delay was due to light's travel time across space – and made possible a rough measure of its speed.

These few examples indicate how technological application and scientific discovery were inextricably linked; each drove and enhanced the other. The notion of a 'pure' versus an 'applied' science does not apply in the 17th century – if it applies anywhere. Minimizing the importance of practical needs – whether military, economic, industrial, medical, or sociopolitical – as the driving force behind developments of the Scientific Revolution would yield an artificial and erroneous depiction of how things really happened.

The linkage of scientific discovery to practical application is perhaps most often associated with Sir Francis Bacon (1561–1626). Born into a well-placed family, educated as a lawyer, elected to Parliament, ennobled as Lord Verulam, and eventually named Lord Chancellor of England (and ousted on bribery charges), Bacon lived most of his life in the halls of power. Accordingly, the topic of power and the building of empire was rarely far from his thoughts. He asserted that natural philosophical knowledge should be *used*; it promised power for the good of mankind and the state. He characterized – or caricatured – the natural philosophy of his day as barren, its methods and goals misguided, its practitioners busy with words but neglecting works. Indeed, although Bacon expressed scepticism of natural magic's metaphysical foundations, he praised magic because it 'proposes to recall natural philosophy from a miscellany of speculations to a magnitude of works'. Natural philosophy should be *operative* not speculative – it should do things, make things, and give human beings power. He considered printing, the compass, and gunpowder – all technological achievements – to have been the most transformative forces in human history. As a result, Bacon called for nothing less than a 'total reconstruction of sciences, arts, and all human knowledge'.

Methodology is crucial to Bacon's desired reform. He advocated the compilation of 'natural histories', vast collections of observations of phenomena whether spontaneously occurring or the result of human experimentation, what he called forcing nature out of her usual course. After sufficient raw materials had been collected, natural philosophers could fit them together to formulate increasingly universal principles by a process of induction. The key was to avoid premature theorizing, navel-gazing speculations, and the building of grand explanatory systems. Once the more general principles of nature had been uncovered, they should then be used productively. Yet Bacon did not advocate a crass utilitarianism. Experiments were useful not only when they produced fruit (practical application) but also when they brought light to the mind. True knowledge of nature served both for 'the glory of the Creator and the relief of man's estate'. While Bacon is clear that one goal of his enterprise is to strengthen and expand Britain – although neither Elizabeth I nor James I responded to his petitions for state support of his ideas for reform – on a larger scale Bacon saw the goal of such operative knowledge as to regain the power and human dominion over nature bestowed by God in Genesis, but lost with Adam's Fall.

Crucially, Bacon considered not only the methods and goals of natural philosophy but also its institutional and social structure. He insisted that older ideals of solitary scholarship had to be replaced with cooperative, communal activity. Indeed, his programme of fact-collecting would require enormous labours, and although he embarked upon such collections himself, he was able to complete very little. Towards the end of his life, he cast his vision of reformed natural philosophy, and the improved society it could create, into a Utopian fable entitled *The New Atlantis* (1626). The story describes the island of Bensalem, a peaceful, tolerant, self-sufficient, Christian kingdom in the Pacific. The happy state of this island is due not only to wise kingship, but even more to the work of Solomon's House, a state-sponsored institution for the study of nature devoted to 'the knowledge of causes and the secret motions of things; and the enlarging of the bounds of Human Empire, to the effecting of all things possible'. The members of Solomon's House study nature communally, although with division of labour and hierarchical arrangement – lower levels collect materials, middle levels experiment and direct, and the highest levels interpret. In Bensalem, Baconian natural philosophers form an honoured and privileged social class, supported by the government, and in service to state and society. Bacon's vision proved inspirational for many 17th-century natural philosophers across Europe as they negotiated their own shifting positions within society.

The rise of scientific societies

Today, scientific research takes place at many sites, some of which even bear resemblances to one or more features of Solomon's House. Scientists work in universities, in governmental, industrial, and independent laboratories, at sites of large and unique instruments (like telescopes or particle accelerators), out in the field or at research stations and outposts, in zoos, museums, and elsewhere. Individual scientists are bound together into social groups by professional organizations, scientific societies and academies, research teams, correspondence, and most recently, the Internet. Funding for scientific research comes from government research grants, corporate research and development, universities, and private philantropy. These three features – physical place, social space, and patronage – are essential to the functioning of modern science. The establishment of these features during the Scientific Revolution was essential for constructing the world of science we know today. Throughout the 17th century and into the 18th, the work of natural philosophers became increasingly formalized. Individuals banded together into private associations which in turn evolved into national academies of science. Individual exchanges of information by letter grew into printed journals. Self-funded amateur and university-based natural philosophers were joined by the first salaried professionals.

During the late Middle Ages, natural philosophical inquiry took place predominantly in universities, monastic settings, and – to a much lesser extent – a few princely courts. These traditional loci of activity remained important during the 16th and 17th centuries, but were supplemented by new venues. Essential to the humanist movement of the Renaissance was the establishment of learned circles of scholarship outside the universities. Within these circles, scholars shared their work with like-minded individuals, receiving support, critique, and recognition as well as occasional patronage. These early groups were mostly literary or philosophical in character. By the late 16th century, however, natural philosophers had expanded the model, thus giving rise to the first scientific societies. The earliest such societies were established in Italy, where dozens were founded in the 17th century – more than anywhere else in Europe. Most of them, however, remained local and short-lived.

One of the earliest societies was the Accademia dei Lincei (Academy of Lynxes). Its name alludes to the emblematic character of the lynx as sharp-eyed and perceptive. The Academy was founded in Rome in 1603 by Prince Federico Cesi – then an 18-year-old Roman nobleman – and three companions, and functioned for about 30 years. Cesi founded his academy upon the belief that the investigation of nature was a complex and laborious affair that required group effort. There were never more than a handful of Linceans, but they included the advocate of natural magic Giambattista Della Porta, Galileo, Niels Stensen, and Johann Schreck, later a Jesuit missionary who brought European scientific knowledge to China. The Linceans pursued projects in all branches of natural philosophy, often independently but occasionally collaboratively, such as their long-term endeavour to publish the *Treasury of Medicines from New Spain* (1651) from manuscripts of Francisco Hernández's expedition to Mexico that had been brought to Italy from Spain. The Linceans embraced the new chemical approaches to medicine, promoted Galileo's work (his 1613 *Sunspot Letters* and 1623 *Assayer* were published under the auspices of the Lincei), and performed microscope studies. Cesi's early death in 1630 robbed the Lincei of its leader and patron, and precipitated its collapse.

In 1657, the Accademia del Cimento was founded at the Medici court in Florence, due in large part to Prince Leopoldo de' Medici's personal interests in natural philosophy. Its motto *Provando e reprovando* ('By testing and retesting') encapsulates the group's focus on performing experiments. The Medici court provided a central locale for communal studies, something the

Lincei had lacked, and Medici patronage kept it running financially. Many members were followers of Galileo, and the group continued several of his research projects and methods. Nevertheless, members of this Florentine Academy worked on everything from anatomy and the life sciences to mathematics and astronomy, and paid special attention to studies and improvements of new instruments such as the barometer and thermometer, in which Leopoldo himself participated. The work of Redi, Malpighi, Borelli, and of many other notable Italian natural philosophers, was carried out within the Cimento. Disagreements between members, the departure of several luminaries, and Leopoldo's nomination as a cardinal which required him to spend more time in Rome, led to the Cimento's closure in 1667. In its decade of existence, the Cimento established the most visible exemplar of a voluntary association of natural philosophers devoting themselves communally to the experimental investigation of nature.

By mid-century, scientific societies spread north of the Alps. In 1652, four physicians in Germany formed the *Academia naturae curiosorum*. Throughout its early years, this 'Academy of Inquirers into Nature' focused mostly on medical and chemical topics. The academy's statutes, published in 1662, declare as its goals 'the glory of God, the enlightenment of the art of healing, and the benefit resulting therefrom for our fellow men'. It grew rapidly, and although members lived widely dispersed throughout German-speaking lands and thus could not meet regularly as a corporate body, the Academy served to link them together virtually, especially through the annual publication (beginning in 1672) of a volume of collected papers submitted by the members. In 1677, Holy Roman Emperor Leopold I gave it official recognition. The 17th-century foundation expanded well beyond medical and life sciences in succeeding years, and eventually developed into the present-day German National Academy of Sciences Leopoldina.

At Oxford University in the 1650s, a group known simply as the 'Experimental Philosophy Club' began meeting at Wadham College to discuss natural philosophy, experiment with mechanical devices, and observe dissections and demonstrations. Christopher Wren and Robert Hooke were early members, and they were joined by Robert Boyle and other notables of mid-century England. Following the Restoration of Charles II in 1660, several members of this Club joined with others to draw up statutes for a more formal corporate organization, and received royal charter in 1662 as the Royal Society of London for the Improvement of Natural Knowledge. The Royal Society, in continuous existence to the present day, marks a new stage in the evolution of scientific societies. Like the Cimento (with which it maintained correspondence), the communal performance of experiments was central, but the Royal Society was envisioned as a much larger, more formalized organization. Over 200 Fellows were soon elected, although most choices among the English nobility reflected wishful thinking about financial rather than any intellectual contributions. Explicitly taking Bacon and his prescriptions as their model, the Royal Society envisioned public and social aims for itself. Indeed, the Royal Society can be seen as an attempt to realize Solomon's House. Many of the early Fellows had been involved in Utopian, educational, and entrepreneurial schemes during the civil war years, and brought these aims to the Society. They strictly avoided sectarian and political attachments, hoping to find in natural philosophy a basis for agreement that could overcome the factionalization of the civil war years immediately preceding.

The Fellows held regular meetings at Gresham College in London, where experiments were performed and new results and observations presented. Virtually every notable British natural philosopher of the period (and since) was a Fellow. The Society's membership soon reached beyond British borders, and election as a Fellow, then as now, carried substantial prestige. Perhaps the most important innovation linked with its early years was the establishment in 1665 of the *Philosophical Transactions*, the first scientific journal, by the Society's secretary Henry Oldenburg. Started initially as Oldenburg's private endeavour – he vainly hoped to earn a living from subscriptions – the *Transactions* soon became conceptually linked with the Royal Society although formally so only later. Oldenburg maintained a vast correspondence (as a result of which he was once imprisoned in the Tower as a spy), and could thus report scientific goings-on across Europe. The *Philosophical Transactions* published not only the Royal Society's activities, but reports and scientific letters from abroad, as well as book reviews. Despite publication predominantly in English, it became a crucial organ for European scientific life – a place to publish observations, announce findings, establish priority, and conduct disputes. Newton's papers on light, optics, and his new telescope appeared there, as did van Leeuwenhoek's microscopal observations mailed in from Holland, and Malpighi's anatomical studies sent from Italy. Arguments over comets jostled for room with reports of monstrous births, and issues appeared whenever Boyle had something relatively brief to report.

Despite its ambitions, the Royal Society suffered the problems common to early scientific societies – loss of key members, financial woes, and lack of patronage. Many of its grand schemes came to naught as a result. A majority of Fellows were inactive and paid their dues sporadically or not at all, and the Crown's sole gift to the Society was the adjective 'Royal'. Its Baconian project for the improvement of trades floundered on the understandable unwillingness of tradesmen to share their proprietary expertise. The English response outside of natural philosophical circles was no better – the Society, its Fellows and activities, were lampooned on the stage in Thomas Shadwell's *Virtuoso* (1676) and their claims to public utility acidly parodied by the 'Voyage to Laputa' in Jonathan Swift's *Gulliver's Travels* (1726). Oldenburg's death in 1677 caused the *Philosophical Transactions* to lapse for a time, and Boyle's in 1691 meant the loss of the Society's most active and generous Fellow. Newton, a Fellow since 1672, became President of the Society in

1703, by which time he was recognized as England's pre-eminent natural philosopher. His prestige breathed new life into the organization, but his tendency to favour work that promoted his own narrowed the former breadth of the Society's activities. Nevertheless, the Society became securely established by the middle of the 18th century, and has carried on ever since.

Unlike the Royal Society that was established from the bottom up, the Parisian Académie Royale des Sciences was established from the top down. It was the brainchild of Jean-Baptiste Colbert (1618–83), finance minister to Louis XIV. Colbert intended both to add glory to the Sun King as patron of arts and sciences, and to centralize scientific activity in ways useful to the state – part of the larger centralization of France that characterized Louis' long reign. The Académie held its first meeting in 1666, with twenty academicians headed by Christiaan Huygens, who had been recruited from Holland. They met twice a week at the King's Library, were expected to work communally (this did not always go smoothly), and received a salary and research support. The French thus realized Bacon's vision far better than did his countrymen. In return for royal funding, academicians were expected to find scientific solutions to state problems. It is no coincidence that the two most highly paid members – Huygens and Cassini – were brought to France while working on the crucial problem of longitude. Academicians also tested water quality at Versailles and throughout France, evaluated new projects and inventions, examined books and patents, solved technical problems at the royal printing press and elsewhere, and produced the first accurate survey of France. The last enterprise, by finding France to be smaller than previously thought, is said to have led Louis XIV to quip that his own academicians had succeeded, where all his enemies had failed, in diminishing the size of his kingdom. Despite service to the state, however, academicians had plenty of time for other studies, particularly several large communal projects they set for themselves, including exhaustive natural histories of plants and animals (Figure 17).

17. A dissection carried out by members of the Parisian Royal Academy of Sciences. The secretary (Jean-Baptiste Duhamel) records the observations while groups of academicians discuss them; the Jardin du Roi (King's Garden) is visible out of the window. *Mémoires pour servir à l'histoire des animaux.* (The Hague, 1731; originally published Paris, 1671)

Royal patronage also provided academicians with workspaces: a chemical laboratory, a botanical garden, and an astronomical observatory on the (then) outskirts of Paris. Completed in 1672, the Observatory of Paris was at first intended as a home for the entire Academy, but became the exclusive domain of the astronomers. The astronomer Gian Domenico Cassini, enticed away from the Pope's service to Paris by a huge stipend and control of the new observatory, took up residence there before the building was finished. He, and three generations of Cassinis after him, made the Observatoire the premier astronomical institution in Europe. Its north–south centreline marked the Earth's prime meridian from which longitude was widely measured for two centuries, until primacy was captured in 1884 by the line passing through Greenwich. (The Royal Observatory at Greenwich had been founded in 1675, shortly after the Paris Observatory, specifically for the 'finding out of the longitude of places for perfecting navigation and astronomy'.) Royal funding also allowed the Paris Academy to send scientific expeditions abroad – to Guyana, Nova Scotia, and Denmark for astronomical observations, to Greece and the Levant for collecting botanical specimens, and famously, in the early 18th century, to South America and Lapland to make observations and measurements to test Cartesian and Newtonian predictions for the exact shape of the Earth. It likewise collected and published observations sent by Jesuits from Siam, China, and elsewhere, and corresponded extensively with members of the Royal Society (even when France and England were at war) and other savants throughout Europe.

Scientific groups beyond the academies

Scientific academies proliferated after 1700, opening in Bologna, Uppsala, Berlin, St Petersburg, French provincial centres, and even at Philadelphia in the North American colonies, and became symbols of national pride and achievement. But academies were only one expression of the developing world of science. They were accompanied by more informal, but sometimes no less important, social groupings. In Paris, the Académie Royale followed upon natural philosophical *salons* held in private homes or public settings, where interested persons assembled for discussion, conversation, and debate under the leadership of an organizer. Their establishment testifies to how developments in natural philosophy had captured public attention, and was becoming a social phenomenon. In London, the new coffeehouses that opened in the later 17th century provided locales for a variety of people to meet and discuss issues, including natural philosophical ones. Public interest fuelled the emergence in the early 18th century of the public demonstrator, a character part natural philosopher and part showman who entertained and educated public gatherings (for an admission fee) with exotic apparatus or showy displays.

Less visible than the academies, but equally significant for the history of science were the networks of correspondence that linked individuals into webs of communication. Natural philosophers privately exchanged letters, manuscripts, and their newly printed books. The privacy of the letter allowed for the airing of unpopular and radically novel ideas, creating a mostly hidden discussion that carried on across Europe throughout the 17th century. This invisible 'republic of letters' (a Renaissance humanist phrase) united like-minded thinkers across national, linguistic, and confessional lines, and bridged the distances between them. The construction of such webs of correspondence was enhanced by people known as intelligencers. They received letters, organized and compiled the information, distributed it to interested parties, and sent out follow-up inquiries. The volume of a busy intelligencer's correspondence could be staggering. Nicolas-Claude Fabri de Peiresc (1580–1637), who encouraged Gassendi and spread Galileo's ideas in France, maintained about 500 correspondents and left behind over 10,000 letters. One of his correspondents, the Minim friar Marin Mersenne (1588–1648), was himself a communications hub. In his monastic cell in Paris, he received correspondence and disseminated the work of Descartes, Galileo, and others through a network across Europe.

In England, Samuel Hartlib (c. 1600–62), a Prussian refugee from the Thirty Years War, maintained correspondence linking all of Protestant Europe and North America; his 2,000 surviving letters are but a small fraction of what he wrote. Hartlib was motivated by Utopian and utilitarian ideas for the reform of education, agriculture, and industry after a Baconian fashion, but also by religious beliefs, particularly millenarian hopes for creating a Protestant 'paradise on earth' in England. His circle included entrepreneurs, moralists, natural philosophers, theologians, and engineers, and his projects ranged from opening technical colleges to improving brewing. The academies themselves became nodes in this epistolary network, and the learned journals – the *Philosophical Transactions*, the *Journal des Sçavans*, as well as their modern descendants – can be seen as formalized versions of it crystallized in print.

Thanks to the establishment of scientific academies and the increasing importance of technological applications in the 17th century, succeeding centuries saw a gradual professionalization of scientific work and a slow disappearance of the 'amateur' natural philosopher. Increased demand for knowledgeable, trustworthy people who could apply scientific knowledge and methods to practical problems drove the establishment of more formal and rigorous training for them in universities, and this in turn led to greater standardization of ideas and approaches. The cumulative result was the 19th-century emergence of 'science' as a career, of 'scientists' as a distinct social and vocational class (resembling in some respects what Bacon had described in the *New Atlantis*), and the gradual refashioning of the early modern world into the modern world of science and technology. That transformation was a slow and complex process, the account of which does not belong in this book. The turns upon the path that historical characters chose, the ideas and needs that influenced their decisions, and the events that enabled or disabled their intentions, were neither obvious nor preordained. While the realities of the natural world would be no different, the ways human beings express, conceptualize, and deploy them might very well be. The particular historical route we have chosen to tread has delivered us into a world of science and technology full of wonders to astonish the greatest advocates of *magia naturalis* and yet not without problems, both those remaining unsolved and those of our own making. Amid our enviable store of natural knowledge, the wise, peaceable, and orderly Bensalem continues to elude us, even if it has never ceased to inspire.

Epilogue

Virtually every text and artefact that has come down to us from early modern natural philosophers expresses their fervour to explore, invent, preserve, measure, collect, organize, and learn. Their innumerable theories, explanations, and world systems that jostled for recognition and acceptance met with various fates. Many early modern concepts and discoveries – Copernicus's heliocentrism, Harvey's circulation of the blood, Newton's inverse-square law of gravitation – constitute the foundations of our modern understanding of the world. Other ideas, like notions of atomism and estimates of the size of the universe, have been incrementally updated and refined by subsequent scientific work, and some, like Descartes' vortices or the mechanical explanation of magnetic attraction, have been discarded entirely.

Modern science continues to pursue many of the questions and aims of early modern natural philosophers – some of which they inherited from the Middle Ages, or even from the ancients. Like Gassendi, Descartes, and van Helmont, modern physicists continue to search for the ultimate particles of matter, to understand how these invisible bits of the universe unite and interact to form the world. Like Kepler, Cassini, and Riccioli, modern astronomers continue to scan and map the heavens, finding new objects and phenomena with instruments far more diverse and powerful than the quadrants and telescopes of Tycho, Galileo, or Hevelius. The explorers in New Spain like Hernández and Da Costa have heirs in scientists who continue to seek for new medicines in the plants and animals of jungles and deserts, or for new life forms in dark ocean trenches and even on distant worlds. Like their Paracelsian and chrysopoeian forebears, modern chemists labour to modify and improve natural substances and to create new materials, continuing the aspirations of Boyle to understand material change and of Bacon to provide things useful for human life. Like Vesalius, Malpighi, and Leeuwenhoek, modern biologists and physicians explore animal and human bodies with new instruments, uncovering ever finer structures and more astonishing mechanisms. Every new electronic gizmo that appears on the market refreshes the ties of technology to the wondrous and the magical.

Alongside such links of continuity, much has changed as well. The deep religious and devotional incentive that motivated early modern natural philosophers to study the Book of Nature – to find the Creator reflected in the creation – no longer provides a major driving force for scientific research. The constant awareness of history, of being part of a long and cumulative tradition of inquirers into nature, has been largely lost. Few scientists today would do as Kepler did when he subtitled his Copernican textbook 'a supplement to Aristotle', or seek for answers in ancient texts, where Newton sought for gravity's cause. The vision of a tightly interconnected cosmos has been fractured by the abandonment of questions of meaning and purpose, by narrowed perspectives and aims, and by a preference for a literalism ill-equipped to comprehend the analogy and metaphor fundamental to early modern thought. The natural philosopher and his broad scope of thought, activity, experience, and expertise has been supplanted by the professionalized, specialized, and technical scientist. The result is a scientific domain disconnected from the broader vistas of human culture and existence. It is impossible not to think ourselves the poorer for the loss of the comprehensive early modern vision, even while we are bound to acknowledge that modern scientific and technological development has enriched us with an astonishing level of material and intellectual wealth.

The Scientific Revolution was a period of both continuity and change, of innovation as well as tradition. The practitioners of early modern natural philosophy came from every part of Europe, every religious confession, every social background, and ranged from provocative innovators to cautious traditionalists. These disparate characters together contributed to the establishment of bodies of knowledge, institutions, and methodologies foundational to today's global world of science – a world that touches every living human being. We could tell them many things they were desperate to know, and they could perhaps in turn tell us things we are desperate to hear. Their age strikes us as both familiar and alien, simultaneously like our own and strikingly different. The very complexity and exuberance of the early modern period renders it the most fascinating and most important era in the entire history of science.

References

Chapter 1

Edward Grant, *The Foundations of Modern Science in the Middle Ages* (Cambridge: Cambridge University Press, 1996), p. 174.

Chapter 2

Giambattista della Porta, *Natural Magick* (London, 1658; reprint edn. New York: Basic Books, 1957), pp. 1–4.

Chapter 3

Nicholas Copernicus, *De revolutionibus* (Nuremberg, 1543), Schönberg's letter, fol. ii*r*; God's artisanship, fol. iii*v*; Osiander's 'preface', fols. i*v*–ii*r* (my translations). A full English translation is Copernicus, *On the Revolutions*, tr. Edward Rosen (Baltimore: Johns Hopkins University Press, 1992).

J. E. McGuire and P. M. Rattansi, 'Newton and the "Pipes of Pan"', *Notes and Records of the Royal Society*, 21 (1966): 108–43, on p. 126.

Chapter 4

Athanasius Kircher, *Mundus subterraneus* (Amsterdam, 1665), preface.

Galileo Galilei, *Il Saggiatore* [*The Assayer*], in *The Controversy on the Comets of 1618* (Philadelphia: University of Pennsylvania Press, 1960), pp. 183–4.

Chapter 6

The Works of Francis Bacon, ed. James Spedding, Robert L. Ellis, and Douglas D. Heath, 14 vols (London: 1857–74), 4:8, 3:294, 3:164.

Further reading

There are several good books surveying the Scientific Revolution in greater detail than is possible here. These include Peter Dear, *Revolutionizing the Sciences: European Knowledge and Its Ambitions, 1500–1700*, 2nd edn. (Princeton: Princeton University Press, 2009); John Henry, *The Scientific Revolution and the Origins of Modern Science*, 2nd edn. (Basingstoke: Palgrave, 2002); and Margaret J. Osler, *Reconfiguring the World: Nature, God, and Human Understanding from the Middle Ages to Early Modern Europe* (Baltimore: Johns Hopkins University Press, 2010). The last is especially good in providing technical details of early modern scientific ideas. A useful reference source is Wilbur Applebaum's *Encyclopedia of the Scientific Revolution* (New York: Garland, 2000), full of short, authoritative articles on hundreds of subjects.

Chapter 1

For the medieval (and ancient) background, see David C. Lindberg, *The Beginnings of Western Science*, 2nd edn. (Chicago: University of Chicago Press, 2007), and for a fascinating account of medieval voyages, see J. R. S. Phillips, *The Medieval Expansion of Europe*, 2nd edn. (Oxford: Clarendon Press, 1998). For Renaissance humanisms, see Anthony Grafton with April Shelford and Nancy Siraisi, *New Worlds, Ancient Texts: The Power of Tradition and the Shock of Discovery* (Cambridge, MA: Harvard University Press, 1992); and Jill Kraye (ed.), *Cambridge Companion to Renaissance Humanism* (Cambridge: Cambridge University Press, 1999).

On other issues in this chapter, see Elizabeth Eisenstein,

The Printing Press as an Agent of Change (Cambridge: Cambridge University Press, 1979); Peter Marshall, *The Reformation: A Very Short Introduction* (Oxford: Oxford University Press, 2009); and Anthony Pagden, *European Encounters with the New World from the Renaissance to Romanticism* (New Haven: Yale University Press, 1993).

Chapter 2

On natural magic and its place in the history of science, see John Henry, 'The Fragmentation of Renaissance Occultism and the Decline of Magic', *History of Science*, 46 (2008): 1–48. On the background to the connected worldview, see Brian Copenhaver 'Natural Magic, Hermetism, and Occultism in Early Modern Science', pp. 261–301 in David C. Lindberg and Robert S. Westman (eds.), *Reappraisals of the Scientific Revolution* (Cambridge: Cambridge University Press, 1990). For an account of various sorts of *magia*, see D. P. Walker, *Spiritual and Demonic Magic: Ficino to Campanella* (University Park, PA: Pennsylvania State University Press, 1995). To correct widely held modern prejudices about the role of religion in science, see the very readable essays in Ronald Numbers (ed.), *Galileo Goes to Jail and Other Myths about Science and Religion* (Cambridge, MA: Harvard University Press, 2009), and for more in-depth treatments, David C. Lindberg and Ronald L. Numbers (eds.), *God and Nature: Historical Essays on the Encounter Between Christianity and Science* (Berkeley, CA: University of California Press, 1989).

Chapter 3

On the major characters discussed in this chapter, see Victor E. Thoren, *The Lord of Uraniborg: A Biography of Tycho Brahe* (Cambridge: Cambridge University Press, 1990); Maurice Finocchiaro (ed.), *The Essential Galileo* (Indianapolis, IN: Hackett, 2008); John Cottingham (ed.), *Cambridge Companion to Descartes* (Cambridge: Cambridge University Press, 1992); Richard S. Westfall, *The Life of Isaac Newton* (Cambridge: Cambridge University Press, 1994). For the best overview of the current understanding of 'Galileo and the Church', see the introduction to Finocchiaro, *The Galileo Affair* (Berkeley, CA: University of California Press, 1989). On astrology, see Anthony

Grafton, *Cardano's Cosmos: The World and Works of a Renaissance Astrologer* (Cambridge, MA: Harvard University Press, 1999). For better understanding of astronomical models and theories, see Michael J. Crowe, *Theories of the World: From Antiquity to the Copernican Revolution*, 2nd edn. (New York: Dover, 2001), and visit 'Ancient Planetary Model Animations' at http://people.sc.fsu.edu/~dduke/models.htm; created by Professor David Duke at Florida State University – this site contains outstanding animations of various planetary systems.

Chapter 4

For Galileo and motion, see the suggestions for Chapter 3.

For other major figures mentioned, see Alan Cutler (for Steno), *The Seashell on Mountaintop* (New York: Penguin, 2003); Paula Findlen (ed.), *Athanasius Kircher: The Last Man Who Knew Everything* (New York: Routledge, 2004); and Michael Hunter, *Robert Boyle: Between God and Science* (New Haven: Yale University Press, 2009). For alchemy and its importance, see Lawrence M. Principe, *The Secrets of Alchemy* (Chicago: Chicago University Press, 2011) and William R. Newman, *Atoms and Alchemy: Chymistry and the Experimental Origins of the Scientific Revolution* (Chicago: Chicago University Press, 2006). For a useful, but now rather dated, overview of the mechanical philosophy, see the relevant sections in Richard S. Westfall, *The Construction of Modern Science: Mechanisms and Mechanics* (Cambridge: Cambridge University Press, 1971).

Chapter 5

Nancy G. Siraisi, *Medieval and Early Renaissance Medicine* (Chicago: University of Chicago Press, 1990) and Roger French, *William Harvey's Natural Philosophy* (Cambridge: Cambridge University Press 1994). On natural history, see William B. Ashworth, 'Natural History and the Emblematic Worldview', in David C. Lindberg and Robert S. Westman (eds.), *Reappraisals of the Scientific Revolution* (Cambridge: Cambridge University Press, 1990), pp. 303–32; and Nicholas Jardine, James A. Secord, and Emma C. Spary (eds.), *The Cultures of Natural History* (Cambridge: Cambridge University Press, 1995). On the Spanish role, see María M. Portuondo, *Secret Science: Spanish Cosmography and the New World* (Chicago: University of Chicago Press, 2009) and Miguel de Asúa and Roger French, *A New World of Animals: Early Modern Europeans on the Creatures of Iberian America* (Burlington, VT: Ashgate, 2005).

Chapter 6

Pamela O. Long, *Technology, Society, and Culture in Late Medieval and Renaissance Europe, 1300–1600* (Washington, DC: American Historical Association, 2000); Paolo Rossi, *Philosophy, Technology, and the Arts in Early Modern Europe* (New York: Harper and Row, 1970); Markku Peltonen (ed.), *Cambridge Companion to Bacon* (Cambridge: Cambridge University Press, 1996); Lisa Jardine, *Ingenious Pursuits: Building the Scientific Revolution* (New York: Anchor Books, 2000); Marco Beretta, Antonio Clericuzio, and Lawrence M. Principe (eds.), *The Accademia del Cimento and its European Context* (Sagamore Beach, MA: Science History Publications, 2009); Alice Stroup, *A Company of Scientists: Botany, Patronage, and Community at the Seventeenth-Century Parisian Royal Academy of Sciences* (Berkeley, CA: University of California Press, 1990).

“牛津通识读本”已出书目

古典哲学的趣味
人生的意义
文学理论入门
大众经济学
历史之源
设计，无处不在
生活中的心理学
政治的历史与边界
哲学的思与惑
资本主义
美国总统制
海德格尔
我们时代的伦理学
卡夫卡是谁
考古学的过去与未来
天文学简史
社会学的意识
康德
尼采
亚里士多德的世界
西方艺术新论
全球化面面观
简明逻辑学
法哲学：价值与事实
政治哲学与幸福根基
选择理论
后殖民主义与世界格局

福柯
缤纷的语言学
达达和超现实主义
佛学概论
维特根斯坦与哲学
科学哲学
印度哲学祛魅
克尔凯郭尔
科学革命
广告
数学
叔本华
笛卡尔
基督教神学
犹太人与犹太教
现代日本
罗兰·巴特
马基雅维里
全球经济史
进化
性存在
量子理论
牛顿新传
国际移民
哈贝马斯
医学伦理
黑格尔

地球
记忆
法律
中国文学
托克维尔
休谟
分子
法国大革命
民族主义
科幻作品
罗素
美国政党与选举
美国最高法院
纪录片
大萧条与罗斯福新政
领导力
无神论
罗马共和国
美国国会
民主
英格兰文学
现代主义
网络
自闭症
德里达
浪漫主义
批判理论

德国文学
戏剧
腐败
医事法
癌症
植物
法语文学
微观经济学
湖泊
拜占庭
司法心理学
发展
农业
特洛伊战争
巴比伦尼亚
河流
战争与技术
品牌学
数学简史
物理学
行为经济学
计算机科学

儿童心理学
时装
现代拉丁美洲文学
卢梭
隐私
电影音乐
抑郁症
传染病
希腊化时代
知识
环境伦理学
美国革命
元素周期表
人口学
社会心理学
动物
项目管理
美学
管理学
卫星
国际法
计算机

电影
俄罗斯文学
古典文学
大数据
洛克
幸福
免疫系统
银行学
景观设计学
神圣罗马帝国
大流行病
亚历山大大帝
气候
第二次世界大战
中世纪
工业革命
传记
公共管理
社会语言学
物质
学习